THE
TREES
IN MY
FOREST

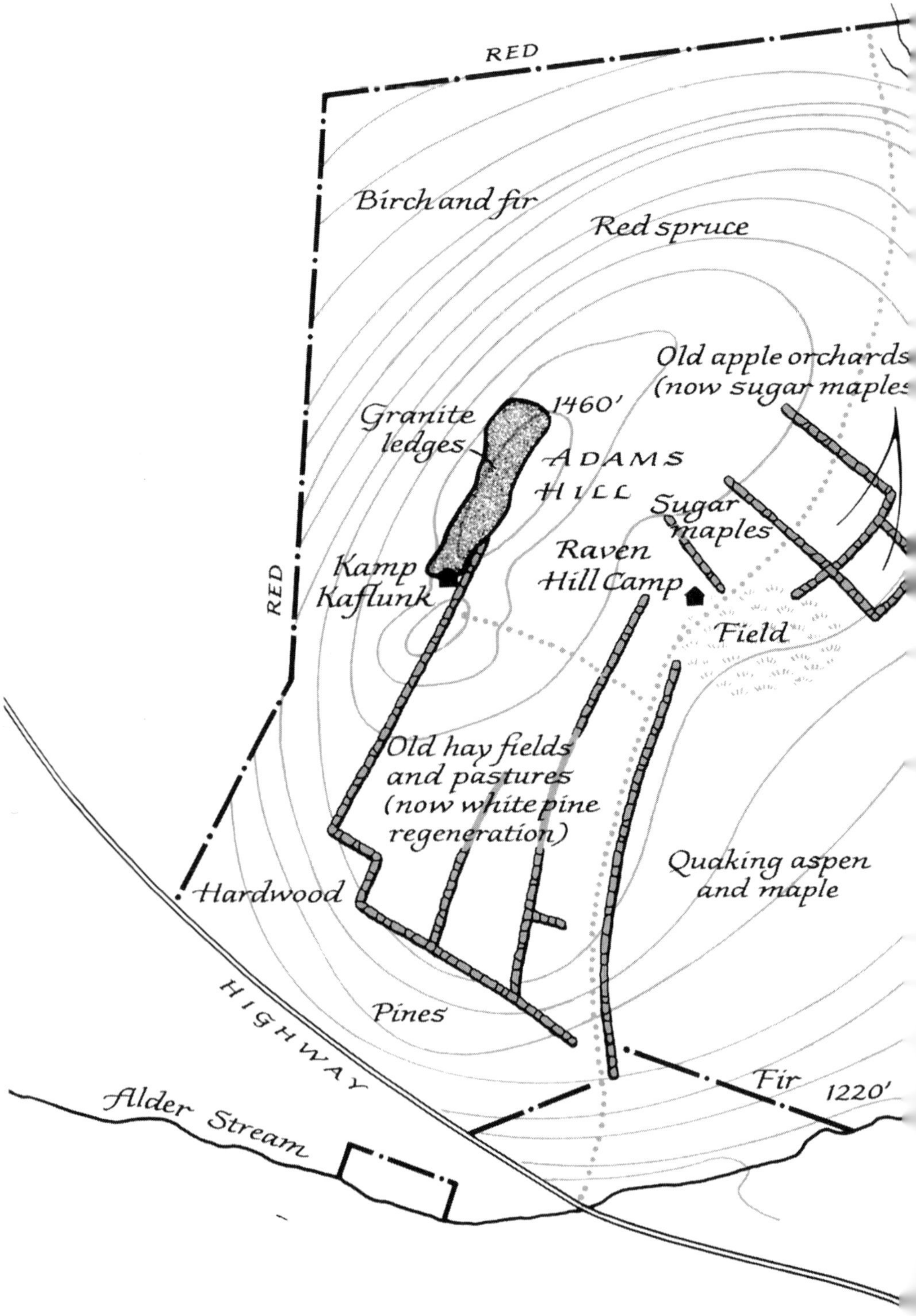

RED

Birch and fir

Red spruce

Old apple orchards
(now sugar maples

Granite
ledges

1460'

ADAMS
HILL

Sugar
maples

Kamp
Kaflunk

Raven
Hill Camp

Field

RED

Old hay fields
and pastures
(now white pine
regeneration)

Quaking aspen
and maple

Hardwood

Pines

HIGHWAY

Fir

1220'

Alder Stream

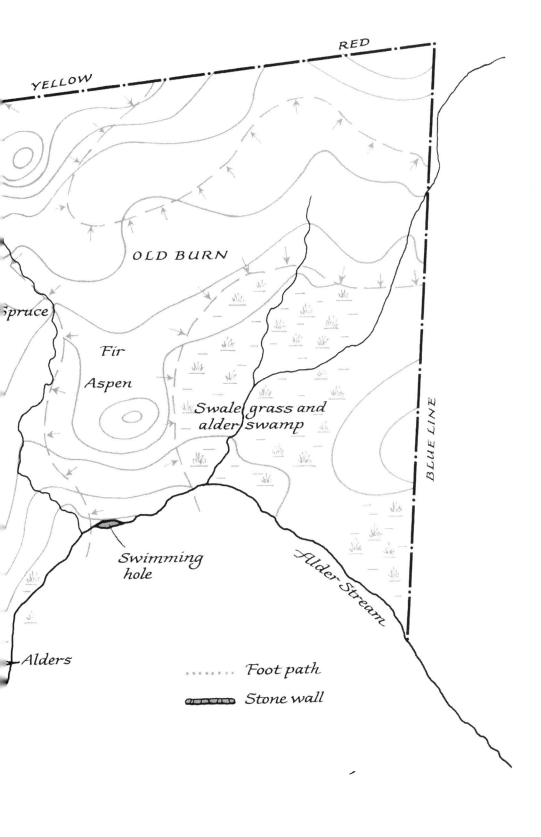

YELLOW

RED

OLD BURN

Spruce

Fir

Aspen

Swale grass and
alder swamp

BLUE LINE

Swimming
hole

Alder Stream

Alders

········· Foot path

Stone wall

Also by Bernd Heinrich

THE TREES IN MY FOREST

BERND HEINRICH

An Imprint of HarperCollinsPublishers

We travel the Milky Way together, trees and men.
—John Muir

Parts of the chapter "Construction for Strength" are modified from an article previously published in *Natural History* magazine.

A hardcover edition of this book was published in 1997 by Cliff Street Books, an imprint of HarperCollins Publishers.

HarperCollins books may be purchased for educational, business, or sales promotional use. For information please write: Special Markets Department, HarperCollins Publishers, Inc., 10 East 53rd Street, New York, NY 10022.

First Ecco paperback published 2003.

Designed by Joseph Rutt

The Library of Congress has catalogued the hardcover edition as follows:

Heinrich, Bernd.
    The trees in my forest / by Bernd Heinrich. — 1st ed.
      p. cm.
    Includes bibliographical references (p.).
    ISBN 0-06-017446-3
    1. Trees—Maine. 2. Forest ecology—Maine. 3. Heinrich, Bernd,
  1940– —Homes and haunts—Maine. I. Title.
  QK164.H45 1997
  577.3'09741—dc21   97-16885

ISBN 0-06-092942-1 (pbk.)

04 05 06 07 RRD 10 9

CONTENTS

Color illustrations follow page 48

ACKNOWLEDGMENTS

This book would probably not have been written if it were not for my Winter Ecology students. Every year, they help me see new things and infect me with new enthusiasm for these woods. I have also walked in the woods with foresters and tried to see the woods through their eyes, always with great pleasure. They are Malcolm "Mac" McClean, Steve Gettle, Steve Pottle, Bob Lesso, Edward Witt, and Cy Balch. I profited greatly from discussions, editorial comments, and valuable suggestions by Bill Alverson, Virginia Barlow, Diane Reverand, John Perry, Rebecca Lowen, and Glenn Booma. Above all I thank my wife, Rachel Smolker, for being a companion, a helper, and a critical editor when I needed one the most.

INTRODUCTION

Having been partly arboreal since the age of eight, I learned early on that trees contain birds' nests, safety, grand vistas, and apples. Climbing tall trees gave me a soaring feeling of achievement. Under the cover of a canopy of green leaves I could peer down, knowing I was safe from the wild boars and other beasts I imagined marauding on the ground. Trees were the best toys for hanging, swinging, daring, and showing off.

For a while I lived on a farm with the family of Floyd and Leona Adams and their three boys, Jimmy, Butchy, and Billy, who were near my age, about ten. At the farm, I was mainly interested in looking for birds' nests, fishing, raising butterflies and moths from caterpillars, and, on occasional nights, chasing after raccoons with the baying hound down in the swamp by Pease Pond. There were rare and secretive birds here and even rarer and more mysterious insects. All lived in the trees in the surrounding forest that seemed to go on forever. The forest was a given, like the air we breathed and the waters of the streams and lakes we swam and fished in. My parents bought a rundown farm with seventy acres of forest just across the hollow from the Adams family, past a forest-bordered pond. In the summer, I ran across the fields and through the shaded hemlock woods between the Adamses' place and ours.

There were heavy snows that first winter, which was also the winter that my parents decided they'd rather work in the woods than in the mill. My father borrowed a crosscut saw from our neighbors Phil and Myrtle Potter. Every dawn, he and my mother, a petite red-headed woman barely five feet tall, trudged through the deep snow into our woods to saw logs for the nearby Rumford paper mill. Everyone else was using chain saws by then. We could not afford one, nor would my father deem it necessary to use one, inasmuch as he determined that the crosscut saw *did* cut logs, given sufficient honest effort. I remember watching them once saw down a large ash tree that, when it fell, crashed out of sight in the snow. They had to dig trenches around it to saw it into the four-foot sections required for paper pulp wood. These they rolled onto a borrowed toboggan, one by one, and pulled them to stack into a neat pile along a path. From there they'd haul the wood out to the main road by sledge pulled by Susie, a light brown horse that belonged to another neighbor, Harold Adams. Meanwhile, I climbed a nearby sugar maple, jumped off, and instantly became buried up to my neck in the heavy, damp snow. My parents had to dig me out, too.

Some years later, I planted a row of trees down our long driveway to the farmstead. I alternated sugar maples, red maples, white ash, red oak, and basswoods. The basswoods hum in June with the thousands of bees that visit their blossoms. The red oaks feed squirrels and blue jays. The white ash seeds sometimes feed flocks of pine grosbeaks, who come from the far north in the winter. The maples give syrup in March, a beautiful display of yellow flowers in May, and spectacular golden and red foliage in October.

We already had other trees around the house. There were two massive elm trees by the barn. One of them was chosen each spring by the bright orange and black northern orioles to suspend their baglike nest, woven out of gray weathered milkweed fibers, onto an inaccessible tip of a long drooping branch. The sugar maples around the house had been planted at least a hundred years before. Some of their thick limbs had broken off, leaving suitable sites where hairy and downy woodpeckers excavated their nest holes. These holes later became nesting cavities for eastern bluebirds and tree swallows. One large cavity from a rotten tree limb was for many years used by a pair of kestrels that hunted mice and grasshoppers in our meadow. These trees were a link to life. Seeing them connected to the birds' lives, I felt somehow connected to them as well.

Life has been likened to a tree of many branches. We and spruce trees are two separate twigs of two major branches. By this analogy, the "seedling" of which the tree of life originated was a bacteria-like microorganism that flourished in the seas and became the ancestor of all life on earth. Our ancestry is still evident in the sequence of bases that make up the genetic code in DNA. The more ancient the divergence from the common ancestor, the greater the change in the sequence of these bases. The amount of molecular change can therefore be used like a clock. We calibrate this molecular clock with known events as seen from the fossil record in dated rocks. According to this molecular clock, life originated about four billion years ago. Plants, animals, and fungi diverged from a common ancestor about one billion years ago. We are all highly evolved organisms, each in our separate ways.

Trees appear vastly different from us, but at the cellular and biochemical levels, they are remarkably similar. Like us, their genetic code is indicated by the same triplets of bases that signify the same amino acids in protein chains. They use the same molecular mechanisms for translating the triplets into proteins. Our biochemical pathways for synthesis and energy conversion are similar. In trees as in humans, genetic information in DNA is packaged in chromosomes in a nucleus and the same mechanisms translate that information into growth through hormones. Despite the vast *potential* alternatives, we find that the closer we look at the organisms of this earth, the more similar (though individually distinct) we are. It is due to common ancestry.

When we take a broader view, we see similarities as well. (But these are probably less arbitrary and more due to convergence.) Trees take in nutrients and respond to their environment. They grow, have sex, reproduce, and senesce. They have predators and protect themselves against them. They succumb to viral, fungal, and bacterial diseases, but they also have defenses. They contend with intense competition from their own kind and others. Although rooted in place, they move, respond to stimuli, and have elaborate mechanisms of dispersing and invading new territory.

Trees are more than distant relatives. They are and have been our intimate associates throughout the whole of our evolutionary history. Our primate ancestors lived in the trees. Millions of years later their descendants built civilizations with trees. The forest provided and still provides us with many raw materials for life, from food to tools, fuel, weapons, clothing, and building materials. Forests ecologically maintain our atmosphere, protect us from flood and erosion, and moderate climate.

I loved trees and forests, and it seemed that I could not find a more appropriate career than forestry, so I enrolled as a forestry major at the University of Maine in Orono. I left school for a year to take part in a zoological expedition into the wilds of Africa. Afterward, I switched my major to zoology. I wandered eventually to California, to continue my study of biology at UCLA, and then to teach in the entomology department at UC Berkeley. When I came back to Maine in 1977, I bought three hundred acres of overgrown farm and recently cutover woodland. It is only an hour's jog from our homestead and the Adamses' farm. The hilly woodland I bought had been a farm very much like the Adamses' farm sixty or more years earlier. Indeed, it was even called Adams Hill, because it had belonged to members of the same family a long time ago.

My purchase of that forest was almost serendipitous. I had asked Mike, a real estate agent and college roommate, to alert me to "a few acres" of land that might be for sale near my favorite hills in western Maine. Loggers here regularly buy land, cut off the biggest trees, then sell cheap to get rid of the land as quickly as possible, to avoid paying taxes.

I could have chosen another place to root my spirits, but fate dictated my choice when Mike called me one day in California to tell me that three hundred acres of logged-over land from Alder Stream and up over the slopes of Adams Hill had come up for sale. The price, about ninety dollars per acre, was too high. Of course, the price was a steal, even then, in 1977. But only if you had cash. All I wanted was an acre or two to set up a shack. I did not want to take out a large mortgage. Luckily Kathryn, my wife at the time, encouraged me to pursue that dream.

Mike, like all good real estate agents, was able to convince us that cash was no problem. Pulling out an aerial photo and drawing imaginary lines to delineate parcels we could sell off to pay for the mortgage, he easily sweet-talked us into buying what we couldn't afford with money we didn't have. I'm no genius at mathematics, but the numbers magically came out *just right*. The next thing I knew we were holding the deed to the property. Ownership forced me to pay taxes and to confront issues of what to do with the land and the potentially emerging forest on it. I didn't realize it at first, but I was also resuming my old interests in forestry, combining them with my love of trees as organisms that are key parts of ecosystems.

Off and on I've lived in forests for most of my life, and I've now owned the forest we'll explore in this book for twenty years. I see it through the eyes of a scientist, an "owner," and a caretaker. I have also used it as my laboratory and classroom. Numerous scientific papers and a few books have resulted from my time in this forest. Over the years, my University of Vermont students have studied squirrel activity, fungal and lichen diversity, tree identification, rodent and shrew populations, animal tracking, overwintering insects, browse preferences of moose and hare, and tree regeneration. I have here studied bumblebee foraging, butterfly and wasp thermoregulation, chickadees' search patterns, raven sociobiology, owl behavior, and tree geometry. I have made many of my best friends in this forest—fellowships born of pleasant time spent wandering its paths and exploring its mysteries together. I have watched this forest change with the seasons and over years. Watched trees sprout, grow, die. . . . Contemplated the long history of human habitation and wondered about its future. The forest of Adams Hill has been my intimate companion and this book, in a sense, is its biography.

A Forest Ramble

I have lately been surveying the Walden woods so extensively and minutely that I now see it mapped in my mind's eye—as indeed, on paper—as so many men's wood-lots . . . I fear this particular dry knowledge may affect my imagination and fancy, that it will not be easy to see so much wildness and native vigor there as formerly. No thicket will seem unexplored now that I know that a stake and stones may be found in it.
—Henry David Thoreau

I'm not much of a hiker of paths, either in a park or elsewhere. Being encumbered like a beast of burden by carrying a pack of goods and tools, and being confined along a trail that leads to some predetermined destination, makes me dig in my heels. I like exploring. I like not knowing when and where I'll end up. That way I get easily diverted and find the new, the unexpected. To learn to know something is less to gaze upon it from known paths and vistas than to walk around it and see it obliquely.

I ramble in my home woods at different times and circumstances. I've struck out in the middle of a blizzard. Once in July I waited until midnight to head out. I've wandered out on spring dawns just when the warblers were returning, on sweltering summer afternoons when the blackflies were biting, in thunderstorms and also under blue sunny skies in Indian summer when the woods were a kaleidoscope of brilliant colors. In my memory, I savor the images that were collected on these rambles and that bind me to this place. Some might consider these images trifles. But I am hard-pressed to come up with greater riches than those memories.

I remember when a blizzard was howling through the great maple trees. A pileated woodpecker, deep black and with its white wing bars flashing, sliced with muscular wingbeats through the forest of thick maples and ash. The back of its head sported a crimson crest. The bird landed abruptly on the trunk of a maple, eyeing me warily. Then it slipped into a cavity in the tree to seek shelter from the driving snow. I remember finding the nests of warblers artfully built and hidden, those of each species in their own special places. I found the nest-cup of a grouse at the foot of a beech. The image of a flock of red crossbills under silent gray skies with great flakes falling on already snow-laden balsam fir trees is imprinted in my mind. I remember a late-summer night under the thick-leaved maples in our woods. The trees let through only points of starlight. It was still. I heard only the faint patter of caterpillar fecal pellets dropping on leaves, and the occasional faint "tseet" of a sleeping bird hidden in the deep layer of leaves. There are moments, other trifles: the crash of a deer through the swale grass; the flushing of a hermit thrush from her nest leaving four startlingly blue eggs at my feet; the black shape of a fisher cat (a weasel relative) disappearing into the brush abandoning a just-killed porcupine with blood on its throat. I do not remember the specifics of innumerable other walks. These walks were perhaps overall very important. They generated the backdrop of familiarity and knowledge that was necessary to make the treasures stand out and to give them substance.

It is a sunny and breezy late morning in August and I'm going on another ramble. It will start, as usual, with the feet, but I do not know where they will carry me. There are always new things to encounter and divert.

Now it seems to be mushrooms. During the last two weeks, the shady ground under the verdure of my young sugar maples had been resplendent with crowds of bright yellow unicorn entoloma mushrooms and a sprinkling of tiny, bright scarlet fairy helmets. There were also brown boletes, purple and green-topped brittlegills, and numerous white-flecked red and yellow death caps. The mushroom diversity seemed phenomenal. In a patch of hemlocks I counted twenty-eight species within fifteen minutes. Every week there is a different mushroom show. These ephemeral fungal fruiting bodies are partially eaten by slugs and squirrels, and they soon subside into the soil as black slime, consumed by fly larvae and bacteria. But the fruiting bodies are only the organisms' "heads." The rest of their bodies are extensive, mostly unseen, and they play a vital role in the forest. Individual lives submerge in this interconnected flow of life.

At this time of year the birds have already receded into the background. Three months ago there was a vibrant cacophony of just-returned birds from all around. I now heard only anonymous "tseeps," the contact calls of hungry young following their parents. Even the red-eyed vireos were silent. In past years, they sang even in early August. Yesterday, I found one of their white birch bark–decorated nests in a horizontal fork of a young maple near the cabin. It was identical to the nest with eggs I had found long ago on the maple at the Adamses' farm, except that this nest was unattended and the one egg

Nest with eggs of the solitary vireo.

it contained was rotten. Are there too few caterpillars for the parents to feed their young this year? There are no indigo buntings this year in the forest clearing around my cabin. A raccoon took all the young tree swallows from my bird boxes. It also took the brood of phoebes from their nest of mud and green moss on the woodshed. A bird's life in this forest is one of boom and bust, and outcomes are often determined by imponderable trifles. As some fail in any one year, others may do well. This year the winter wrens seem to be having great success, perhaps because my recent lumbering operation has left a scattering of limbs and upturned roots under which they like to nest. Families of these little brown birds flit like airborne packs of mice along the ground. Two weeks ago I found four black-throated blue warblers fledged from a nest just a foot above the ground in my dense patch of maple saplings. Then I saw four pinfeathered young of the black-throated green war-

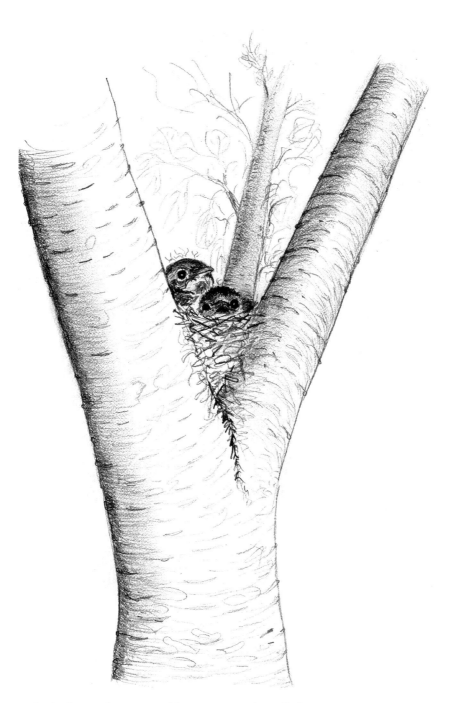

Black-throated green warbler young ready to fledge.

bler huddled in a nest wedged into the fork of a shaded yellow birch in mixed woods by the outhouse. These are but two of the twenty-two species of warblers in this forest on the hill. The nests that I find are just a few of the very many that are there. Like the innumerable fungi that so rarely show themselves, these are visible reminders of hidden life that one seldom sees.

History also lies hidden, but there are hints. Wherever the skidders, a type of logging vehicle, have scratched the topsoil I find small bits of charcoal that jolt my memory. I had in years past found similar bits in scratched soil on nearby Mount Blue and Tumbledown Mountain. Now was the perfect opportunity to search for clues to the fire, because the stumps of the recently cut trees would expose history. I examined the growth ring patterns of dozens of trees, concentrating on the largest red spruce stumps I could find. All were of trees 130 to 155 years old. For all that age, these trees were no more than two feet in diameter. In all of these old spruces (as well as an ash, a cedar, and a white pine) the first 60 to 100 years of their lives showed only microscopic growth, but at varying times after that, the width of the growth rings gradually increased. Even when the trees were cut their growth had not yet decreased, because they had not yet attained their maximum age or size. I inferred that the trees on this land thus have the potential to live and grow for more than 155 years, but the fact that there was probably not one tree on the entire hill older than that suggests that something destroyed all of the trees about 155 years ago. Was the charcoal a clue? Perhaps a massive conflagration swept over the whole countryside and left this and the neighboring hills denuded and charred. After fire had swept through, the land would have been invaded first by the windblown seeds of fireweed. Thickets of poplars and pin cherries followed. The young fast-growing trees would have started out together and soon shaded the ground. Spruce and other shade-tolerant species could then have invaded but they would have grown slowly (hence their initial tiny growth rings) as they were dominated by the poplars for many decades.

The short-lived initial invaders would have died one by one, leaving small canopy openings in which the spruce and other secondary invaders could grow faster, putting on wider growth rings that continued until the present time. Until last winter's lumbering operation, they in turn had inhibited the growth of seedlings under them. Now that there were holes in the canopy, there would once again be a profusion of new red spruce, balsam fir, white pine, red and sugar maple, ash, and beech, reaching for the light.

After reading growth rings on stumps I traveled on along skidder trails. Moose were using these trails, too. In three hours I saw three of them, and two deer. Of the deer, I saw only a half second's flash of brown and white, and heard several crashing leaps through the woods. The first moose was a bull with his antlers in dark brown velvet. He was grazing ahead of me along the logging path. Every time he lowered his head I crept closer. He saw me when I was within ten yards. After looking at me for several seconds, he finally dashed off at a fast high-stepping gangling trot.

Moose are attracted to the vigorously growing hardwood underbrush that follows natural blowdowns and logging operations, including clear-cuts. People who favor massive clear-cutting often claim that moose thrive in clear-cuts. But what they usually *don't* mention is that in some clear-cuts they get rid of the moose browse that would normally grow there. They use helicopters to spray herbicides that kill the regenerating young hardwood trees to culture unpalatable and sterile pine and spruce plantations.

I walk often in my forest and so I'm usually no more aware of

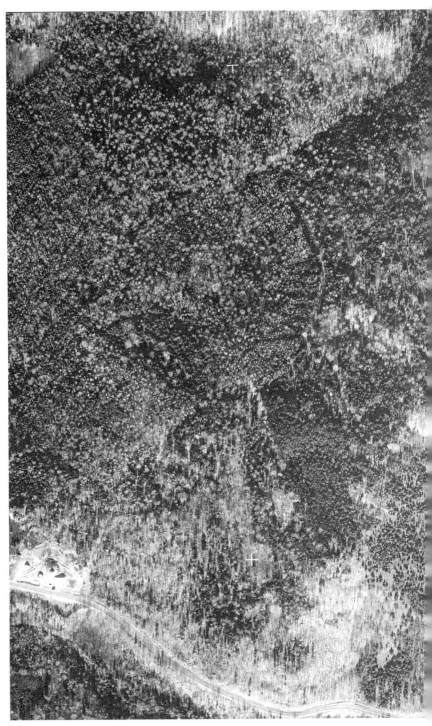

Aerial photographs of my forest in 1966 (above), and again in 1992 (overleaf).

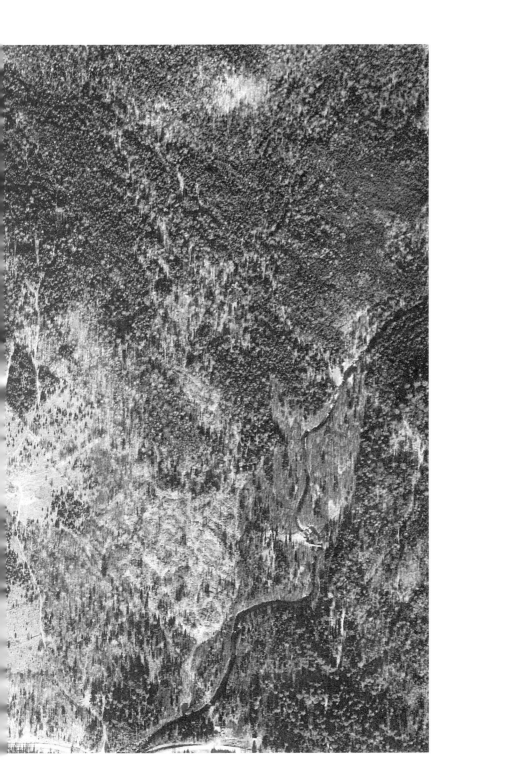

growth in the trees than of growth in Maine's human population. But the opposition of snapshots separated in time can be startling. Thanks to the USDA-FSA-Aerial Photography Field Office in Salt Lake City, Utah, such snapshots are available, and I purchased the earliest (1966) and latest (1992) aerial photographs of the area that includes my forest. The November 20, 1966, photograph showed fully cleared fields to the west of the footpath to the cabin, and in the area where the cabin now stands. The surrounding, stonewall-enclosed field to the north of the cabin, the area that is now my sugar maple grove, looked like a cleared field except for a few scattered young pines. The old field to the west that is now my pine grove showed a gradation from scattered to continuous young pines. There was a clear-cut to the west of the homestead clearing, extending all the way down to Alder Stream. Last year my lumbering crew took out a lot of quaking aspen from this old clear-cut area where these pioneering trees were starting to topple. Balsam firs were lately growing slowly under them. Now the firs will grow fast, and the aspen are regenerating, from their own root stalks, shooting up in thickets and gaining five feet of height per year. This is a prime moose browse area. Grouse will soon also be attracted by new aspen buds and the protection from predators that the dense growth provides.

In the 1992 aerial photograph, all of the land already shows a continuous cover of unbroken forest with the exception of the clearing that I've laboriously maintained around the cabin. My walk today is to the north of the cabin, where there were dense stands of balsam fir. The trees are now falling in swaths. In a several-acre section the tangle of balsam firs was so dense that even the oldest trees were merely thin poles. At this site the loggers cut little. I wished they had taken more.

Regeneration of trees from the seed-leavened earth is phenome-

Indian pipe flower among balsam fir seedlings and hairy cap moss.

nal on cleared ground. In all places where the sunlight is now admitted to the forest floor there is a bloom of trees that have been released from decades of waiting. On months-old skidder trails seedlings grow almost like moss in dense green crowds. Where the forest was cut selectively, the mix of regenerating species is close to that of the original forest cover that grows there already. In one mixed stand of red spruce, balsam fir, hemlock, white birch, white

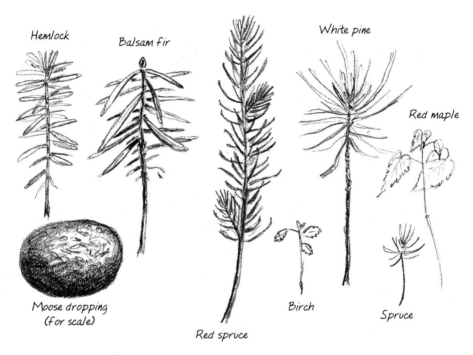

Hemlock

Balsam fir

White pine

Red maple

Moose dropping
(for scale)

Birch

Spruce

Red spruce

Some tree seedlings.

pine, and red maple I counted the following seedlings: fourteen spruce, ten birch, six hemlock, two red maple, one cherry, and one poplar. All were packed into one two-foot-square area. These were by no means maximum numbers. I counted up to eighty inch-tall tree seedlings per square foot. If the environment is suitable, suitable trees will grow there. If it is not, they won't. I do not know which among these potential trees will become a real tree, but in this forest the *land* will select the best one for every square foot of its ground. It has much "choice" in the naturally provided annual rain of native tree seeds, and it can hardly fail to pick the best tree for each slot.

By far the vast majority of seedlings soon die, because each mature tree in this forest requires at least one hundred square feet of ground. By my calculation that leaves 1 tree growing up out of

80 × 100 or 8,000 *potential* trees, and those potential trees for that one slot are only the candidates that came forth in *one* year. For the life of a tree there are new candidates to replace it each year. It is plain to see that planting trees in this forest is akin to sprinkling a lake to keep it wet. It only makes a difference if you first drain it dry. Those largely foreign-owned companies that do just that (through chemical spraying from helicopters to kill trees) brag about all the tree planting they do. They do not mention that for every one tree they plant—and they plant only species selected for their immediate commercial potential—they may have to destroy more than eight thousand seedlings first. These, the eliminated seedlings, contain the best competitors for that area, which is of course precisely why one needs to go to extreme lengths to try to kill them off if one wants to grow something else.

I left the cuttings to come back up to the ridge, reclining on one of the dark green moss-covered boulders left by the glaciers about ten thousand years ago. Boulders now protrude all over my hill. As I lingered under the red spruces on the sun-dappled ground I heard a soft breeze through the branches above me, and I marveled at how this rock pile has been transformed, by trees. At one time, soon after the glaciers, lichens encrusted the rocks. Vines of crowberries, blue-berries, cranberries, and sedges grew on gravelly soil. They held moisture. Mosses took hold, retaining even more moisture. A layer of brown humus accumulated through the ages. Spruce, fir, and birch seedlings sprouted. Vines and roots, building soil and supporting moss, crept over the rocks, building yet more soil. The first small trees grew. Fallen limbs and trunks covered rocks. Now, ten thousand years later, the aerial photograph shows an unbroken forest of tall trees. Where I walk, I still sometimes see the tops of a jumble of rocks reminding me of what once was, but in the spring I hear the chorus of the hermit thrush and the Blackburnian warbler, and they are sounds of coniferous forest.

The stillness in the shade of red spruces and hemlocks in August was only broken by "tseeps." I searched the foliage for a long time before I noticed their source: a family of brown creepers, birds that confine themselves to the trunks of the trees where they search for spiders and insect eggs in bark crevices. The parents hopped in little jerky motions up each tree trunk, all the way into the branches, then they dove straight down like a falling rock to within six inches of the ground onto the trunk of the next tree. Then they repeated the process. They never stopped except to feed their young, who often perched motionless in mid-tree ascent. When the creeper family moved on, I heard the "tseeps" of a family of black-capped chickadees high up in the branches. The soft nasal quacks of a red-breasted nuthatch chimed in. A solitary vireo, one of the first migrants to arrive and sing in the coniferous forest, stopped by, looked me over with black eyes surrounded by shiny white eyerings, made a harsh, grating call, and flew on. I had previously found its nest. As always, it had been built in a young balsam fir close to the ground. There are some who point out that clear-cuts attract birds. They do. Some species. Birds are habitat-specific. Bobolinks and meadowlarks don't live in spruce forests. They live in meadows. The presence of some birds is an indication of forest degradation or of forest loss, not of forest health. In this forest I've counted 105 bird species, and each has slightly different habitat requirements. This means that there is habitat complexity. That is what *makes* it a forest.

More obvious than the songs of warblers and vireos were the signs of woodpeckers. A live red spruce directly next to my perch had five holes typical of those made by pileated woodpeckers. The tree must have had internal decay that attracted carpenter ants. And the bird hammered through solid wood to get at the ants. Yellow wood chips were still strewn on the brown spruce needles and velvety green moss just below the five oblong holes. How did the bird know that ants (totally comatose in winter) were inside? Just to the side of my climbing tree, a red spruce, stands a hemlock. Almost the entire lower south side of this tree was pockmarked with pea-size holes that a yellow-bellied sapsucker had drilled some years ago. (If the holes had been drilled recently, the red color of the hemlock bark would show beneath its brown surface.)

Red spruce are declining in many parts of New England, in part due to acid rain coming from the Midwest. But here on the rocky ridge top, in my forest, they still prosper. Nevertheless, not every area and not every era is ideal for spruce. Every few yards provides a slightly different habitat for a tree, as every few inches of soil is a different habitat for a seedling.

Just a few yards below the rock on which I'm sitting, in a depression in the jumble of glacially deposited granite boulders, lies a dead pin cherry tree that fell last year. A few smaller dead pin cherries and mature gray birch and striped maples of the same age remain nearby. All will fall in five to six years. The presence of these short-lived trees means that thirty to forty years ago there was open ground here, the pasture of the old farm.

New, longer-lived trees are now on the way. Balsam fir and red spruce already one to ten feet in height grow in the shade below the steadily upward-reaching beech and yellow birch. For now, they are barely holding on and growing only two to three inches per year. Should one of the larger trees that now shades them fall, they could take off toward the sun, growing up to two feet per year to quickly fill the void. All the ingenious strategies that different tree species use to tap the incredible amount of solar energy that is available are contingent on what competitors do.

I'm sure the BTU equivalent of energy captured by a growing tree has been calculated to the third decimal point, but to me that figure provides less meaning than the heat I feel when I burn a stick of wood in my cast-iron stove. Two or three dried split pieces of rock maple can make it glow red-hot, warming the stove and the house. Heat is a form of energy. The source of that energy, captured by the trees' leaves, is the sun. Multiplying the potential energy of those two to three pieces of split wood by the untold thousands of logs accumulating in the trees all around me, I am awed by the sheer magnitude of energy that drives life, passing from one form to the next. The energy captured by trees and other plants will eventually be tapped by bacteria and fungi, by insects and other herbivores, and then passed on to birds and other predators, like us.

Given the constant extravagant input of energy into the forest and into life, it is small wonder that the evolution of the most extraordinary complex creatures, as well as human civilizations, has been possible. After looking at trees, and heating coffee on my woodstove, it is not difficult to comprehend how life can proceed toward incredible complexity, such as a hummingbird or a moth, in a seemingly "uphill" direction from chemical chaos. Organization takes energy, and energy on earth comes from the sun through photosynthesis, the chemical process in leaves that is responsible for almost all life on earth. Photosynthesis, the reaction that uses the sun's energy to remove hydrogen atoms from water and to attach them to carbon dioxide from the air to make sugar, is largely due to the miracle of the chlorophyll molecule that colors plants green. It allows trees and all the life forms that depend on trees to use the nuclear fusion reactions of the sun's inferno. Curiously, chlorophyll is a complex ring structure binding a metal ion that is similar to the hemoglobin molecule in our blood. Trees and warm-blooded animals both owe their existence to similar and very intricate molecules that power life. We balance on a ray of light and an oxygen molecule. Functions based on such delicacy are easily disrupted.

The creepers and chickadees left, and even the nuthatch became silent as I continued to sit on my rock. Gradually, the wind picked up. The trees swayed, and the leaves shook above me. Sunspecks flickered wildly on the forest floor in a dance that ebbed and flowed, keeping time with the changing intensity of the wind. For a few minutes, I heard the steady cadence of a long-horned beetle larva chewing in dead wood in the background. An ichneumon wasp with dark blue wings searched on the nearby yellow birch. Flying from one sunlit leaf to the next, the wasp vibrated its two black outstretched white-ringed antennae ahead of it, palpitating for scent trails of caterpillars.

A View from the Top

For beauty, give me trees with the fur on.
    —Henry David Thoreau

My ramble takes me to the top of the red spruce next to the boulder. Carrying a clipboard with paper in one hand made the climb a little slower than normal. I passed through a few spiderwebs built close to the trunk of the tree. A basking mourning cloak butterfly took wing from near the top.

In mid May, when I last climbed this tree, I photographed a forest panorama. Far and wide the poplars and birches were then leafing out, turning a light pea green. The red maples had not yet put out leaves and their gray twigs were highlighted with burgundy blossoms. Flowering shadbush and cherry trees shone in white patches among the light greens of erupting poplar leaves and the dark black greens of evergreen firs and spruces. This year (not the previous nor next), all the sugar maples were resplendent with small lemon yellow flowers dangling loosely at the ends of the twigs. Now, in early August, the vista all around was one of almost uniform green.

The far distance mesmerizes me, but my eyes eventually return to my perch, the two inch-thick limbs near the top of the spruce. I

View of my forest and clearing with cabin from Bald Mountain, January 1990. (Glenn Booma)

am at ease with my left arm around the slender spruce treetop and hold my notepad while writing with my right hand. Three months ago, when I was in the same spot, the tips of the spruce twigs all around this perch had bright red nubbins at their ends. They were flower buds. Those forming male catkins have now already turned into brittle bean-size cones that disintegrate into peppery chaff when I rub them between my fingers. Their pollen has long been shed. The female flowers have grown into firm green full-size cones now holding fertilized seeds that will mature by fall. All of these fresh cones are covered with translucent yellowish gobs of viscous, sticky pitch,

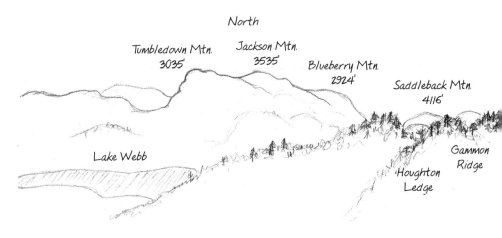

The panorama through 360 degrees from the top of a red spruce.

the stuff of amber. The pitch tastes unpleasantly pungent and is long-lasting on the palate. I suppose that, at least for the moment, it is an effective deterrent to squirrels as well as to me. In the coming winter, however, some red squirrels will snip off the ends of the twigs laden with cones, and after dropping a dozen or so, descend and feed on the ground, hoarding some in larders for future use.

There are very few squirrels in the woods this year. Last year there were no fir cones, no red spruce cones, no acorns, no beech-

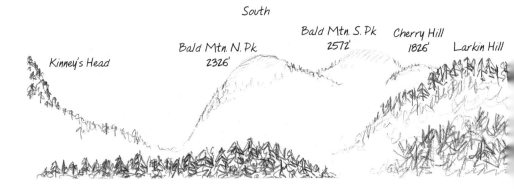

East

Mt. Blue
3187'

Wilder Hill
2000'

Gleason Mtn.
1800'

Kinney's Head

nuts, and no sugar maple seeds. Gray squirrels left their individual haunts all over New England. Their exodus in search of food was evident by the epidemic roadkills that even invited comments in the newspapers far and wide. The highways through New York and New England were littered daily with their carcasses. I kept a tally during a trip on Highway 87 to New York, counting more than two hundred (exact numbers were difficult, because at least fifty were reduced to "smudges"). The mass of squirrels is gone. Perhaps it is not coincidental that the trees are now poised to produce banner seed crops. As I scan the neighboring red spruces, I observe every one resplen-

West

Mt. Washington

Lake Webb

dent with an ample crop of fresh cones. The balsam fir trees have fresh cones as well, and the sugar maples, red oaks, beeches, and beaked hazel are all producing seeds this year. A feast is in store for the fortunate few, and that feast will ultimately have considerable impact on the forest.

As I scan the vast expanse I see the green forest from this grand view fade to a blue haze in the distance where all detail is lost. Nevertheless, a few white pines tower up like black jagged teeth on the ridge tops of Kinney's Head two miles away.

The impression from my spruce, even though it is on the top of a hill, is of being in a huge green bowl. The bowl is formed by Kinney's Head, Gleason Mountain, Wilder Hill, Bald Mountain, Saddleback, Houghton Ledge, Mount Blue, and Tumbledown and Jackson, the mountains that form the major watershed for the Androscoggin and Kennebec Rivers. Solid unbroken forest is all around me, stretching far beyond my vision, for hundreds of miles. It is one of the few such forests remaining in the world. The forest regulates the water flow from the frequent heavy rains. It prevents floods, providing steady runoff into the trout-filled streams. It used to support salmon runs. Such a forest is also the diffuse lung tissue of the earth to which we are irrevocably bound. It is not our "environment." It is us.

WHISPERING PINES?

And this our life exempt from public haunt,
Finds tongues in trees, books in the running brooks,
Sermons in stones and good in everything.
　　　　　—William Shakespeare

The white pines seemed to be one of the only tree species in my woods without a seed crop this year (1996). I wondered why, knowing that pines reputedly produce seed "sporadically." Perhaps I didn't check closely enough. Perhaps the trees I examined were just too young.

White pines grow abundantly along the route I travel frequently between Vermont and Maine. They are visible almost continuously along Routes 142 and 2 all the way to Gorham, New Hampshire. On one of my trips back to Vermont, I casually looked at the trees. No cones here, either. That seemed like corroboration. I was almost convinced that this was not one of those sporadic years that pines flowered or produced seed. However, as Nietzsche said, it is sometimes less important to have convictions than to have the courage to attack them. Similarly, in biology it helps to be paranoically skeptical. Although I continued to be deficient, I still kept looking up at the pines along the highway on my drive to Vermont, even long after passing Gorham. Suddenly, I saw branches of white pines *laden* with cones. And it continued. For miles on end all the pines were conspicuously adorned with long fresh green cones.

It was disconcerting to have worked toward a seemingly well-tested conclusion, namely that all white pines were not flowering this year, only to have it proved wrong down the road. Yet, I felt strangely excited: Now I *really* started to look. A few miles later—again no cones.

The next week I drove through Adirondack Park in New York State. Along Routes 8 and 87 there are almost continuous vistas of white pines. The pattern held. As in Maine, for ten to twenty miles all the mature pine trees were laden with the current year's crop of young green full-size cones. For the next ten to twenty miles the trees were bare. I remarked to my wife-to-be that I had seen pine trees for forty years, and I had never noticed this before. She also had never observed it but now felt that it was "obvious."

We both study animal behavior, and so launched into a discussion of the behavior of pine trees. Though trees aren't animals, the same principles apply. All organisms do what they do and are what they are because of selective pressures operating on survival and reproduction.

There are two main questions. *How* do pine trees time their reproduction in any one area to be synchronous, and if so, *why*? In answer to "why?" there is at least a possible explanation. Pine seeds are highly prized food. They are eaten by crossbills and squirrels. It might be advantageous for a tree to produce seeds at about the same time that neighboring trees do. A pine tree that produces a crop of cones in a year when all of the neighboring trees do *not* will quickly be denuded of cones by hungry birds and squirrels. It is far better for each tree to release its seeds in a large burst just when all the other trees do, because then the seed predators' appetites are more likely to be satiated on the seeds of other trees.

We came up with several plausible alternatives to the *how* question. Then we tried to reject as many of them as we could. We rejected the idea that age explained the phenomenon, although we figured some minimum age is necessary for reproduction. We rejected the idea of a synchronizing signal transmitted through the roots, because the fruiting pine stands were often bisected by roads and streams.

Many plants flower in response to stress, as if to take advantage of one last opportunity. *Seasonal* time of flowering is determined by the relative hours of light and darkness in a twenty-four-hour period (the photoperiod). But neither stress nor photoperiod seemed to determine the behavior of the pines. Maybe they can give a signal that says, in effect, "I'm making flower buds" (as opposed to merely shoot or leaf buds), and this signal, detected by other trees, causes them to do likewise. Pines can't whisper these messages to each other, but they may have other forms of communication. By the process of elimination, we felt that an airborne chemical signal—a pheromone—is possible. Perhaps some trees, like many insects and some mammals, communicate by airborne chemical signals to synchronize their reproduction.

The idea is not unprecedented. For example, it is more than folklore that one apple spoils the whole bunch. A rotting apple releases a gas, ethylene, that acts as a hormone to accelerate maturation (and rotting) in the others. Similarly, Jack C. Schultz has shown that some trees whose leaves are injured by caterpillars feeding on them induce neighboring trees to fortify their leaves with toxins that inhibit caterpillars from attacking. In pineapples, ethylene induces flowering. In many other plants it induces leaf and fruit abscission or "cutting off." Given the right context, trees might also signal to each other that they are going to produce seed, encouraging the neighbors to do so too, thereby making it safer for all participants by "flooding the market," providing a glut of food to predators. If so, who goes first? We do not know, because we have not listened closely enough to the language of the trees.

Trees?

*Every tree like every man must decide for itself—will it live in
the alluring forest and struggle to the top where alone is
sunlight or give up the fight and content itself with the shade.*
———Ernest Thompson Seton

In my forest there grows an ancient line of small unobtrusive plants
that remind one of trees and that are probably descendants of trees.
They are commonly called ground pines and ground cedars and are
botanically known as lycopods or club mosses. Because of their evolutionary history, as well as their appearance, they are reminiscent
of trees, and they force closer scrutiny of the common definition of
a tree. The definition "a tall woody plant" is adequate for most lumberjacks, but it is rather limited for the scientist because it defines
no more about true relationships among plants than does color or
body size among mammals. In plants, size may vary enormously
even between very close kin. In my backyard, for example, is a
black willow tree, having a diameter of about three feet and a height
of nearly forty feet. In Alaska, a very close relative of this tree, the
Arctic willow, creeps along the ground as thin tendrils, and the only
height it achieves is in the spring after the snow melts when it lifts

flower catkins all of an inch above the ground. Aside from size, the differences between these two plants are minuscule. Bamboo, saguaros, and bananas are testimony that even grasses, cacti, and herbs can become "trees" just by having their stems strengthened and lengthened.

A whole ground pine plant that is easily buried would not likely convince anyone that it is a tree. Sprigs of club moss, however, have such a close resemblance to conifer that they are commonly used for Christmas decoration as a credible substitute for fir boughs. Seeing the shiny evergreen needles of a ground pine or a ground cedar one might easily confuse them with the twigs of a "real" coniferous tree. In my forest, the twigs of the ground cedar, *Lycopodium complanatum,* for example, have scaly oppressed leaves like those of cedar or sequoia. Those of the three other species of "ground pines" have sharp needlelike leaves that look uncannily similar to twigs of the giant araucaria trees, known as klinky pines or monkey puzzle trees, native to Australia and South America. We now commonly purchase these holdovers from the days of the dinosaurs at shopping malls and keep them as houseplants.

A plant that superficially looks like a somewhat tall moss may appear to have little claim to the title of "tree." But then consider a tiny fence lizard that for evolutionary reasons reminds us more readily of dinosaurs than does an elephant. So do club mosses remind us of trees, because their ancestors (or the close relatives of their ancestors) once were trees. Both are lycopods. At one time lycopod trees dominated all of the forests of the globe, and they did so for a period spanning at least a hundred million years. The living relatives they left us are now midgets, yet the irony is that they now thrive in an extremely competitive society of giants.

The ancient lycopods included *Lepidodendron,* a tree some 150

Sketch of a piece of Lepidodendron tree twig fossil in rock (natural size).

feet tall. Lepidodendrons were loftier than any trees now in my for-est. Some of them were even taller than the giant white pines. In the late Carboniferous era (Pennsylvanian) forest ferns and horsetails also grew to be trees. A few tree ferns still remain in the tropics, but the lycopods and horsetails that remain are herbs. The local horse-tails are called scouring rush. They still grow here along sandy road-sides and stream banks. Before the invention of Brillo pads, the Adamses on this hill probably used them to scour pans because the horsetails' abrasive silicon ridges are useful in removing kitchen irri-tations such as burnt-on apples and bear grease.

Back in the Carboniferous era, when Lepidodendrons ruled, there were no mammals. No birds flew. There were giant froglike amphibians, six-inch cockroaches, and the monster dragonfly, *Meganeura monyi*. Some of the carbon that the forest mined from the air by photosynthesis with the aid of the sun's energy accumulated in thick mats that were covered and compressed by sediments. These mats become the coal beds that much later fueled the Industrial Revolution. The solar energy trapped in carbon-carbon bonds of those trees is even now still being used by us.

Just as we recognize the resemblance between the wing of, say, a green darner dragonfly, *Anax junius*, and a fossil imprint of the wing of an extinct giant dragonfly, so it is easy to recognize the similarity in body form between a present-day club moss and the fossilized forms of a Lepidodendron tree. Lycopod trees branched, then branched again and again into smaller branches. If woody supporting tissue were incorporated each year into the whole length of a present-day lycopod (rather than only adding inches to the tips), it would end up with a form fairly close to that of a Lepidodendron tree. Nevertheless, their relationship to pines and cedars is obscure.

The lycopods are acknowledged to be among the most morphologically bizarre and, consequently, fascinating group of "lower" vascular plants. Recent molecular data from gene sequencing confirm that they are unique and also among the earliest diverging land plants. That is, although their leaves may look like some present-day conifers, they are not closely related to them. They are still a unique ancient lineage. The fact that they occurred in either tree or herb form is of minor importance in defining relationships, because wood can come and go. It is just one of several means of achieving height in a tree.

Even wood itself has evolved independently several times. Flowering plants, such as oaks and maples, and conifers, such as pine and fir, have different ancestries that are reflected in the anatomies of their wood. The first have long cylindrical vessels consisting of tubular cell walls stacked end to end. The end walls between adjacent cells are digested by the living cells before they die, leaving extremely long tubes through which water rises in the tree. In an oak tree these vessels can be seen with the naked eye. In conifers, in contrast, water is conducted through much smaller, long, spindle-shaped cells called tracheids. Like vessel cells, these also die in the growing tree. However, water percolates from one tracheid to the next through numerous little holes or perforations through the cell walls. In both types of these very different plants that we call "trees," which have very different support structures we lump as "wood," wood is still produced in the same way. It is derived from a single layer of actively dividing cells, the cambium, that is under the trees' outer bark. When these cells divide, the cell layer produced to the inside dies and becomes wood, the water-conducting and structural tissue, while the other (outer) cell of the division becomes phloem, the sap-conducting tissue that ultimately becomes bark.

In Lepidodendron trees, support came not only from wood, but also from bark. The ancestors of today's horsetails also invented wood, but of another kind, and they built trees with it that reached eighty feet high. Ancient horsetails also stayed erect by huge underground stems that interconnected trees. Such underground stems or rhizomes still exist in present-day equisetums that, like our club mosses, no longer produce wood.

There are still other ways that have been invented for reaching high. Cycads, for example, are supported by an external armor of leaf bases and by concentric rings of wood produced by concentric internal cambium layers. These single-stemmed ancient trees dating back some 250 million years still exist, but only as isolated remnant populations in the tropics.

The tree growth habit of a very tall trunk is also found in a number of plant groups that do not produce wood—palms, bamboos, tree ferns, saguaro cacti, some euphorbs. Tree ferns prop themselves up by an external mantle of wiry roots. Palms, bamboos, euphorbs, and cacti use girdling strands of stiffened vascular tissue rather than wood. Palms in addition use external roots and also clasping leaf bases.

Anyone can distinguish a short mosslike herb such as a ground pine or a ground cedar from a "real" pine tree and a cedar without too much trouble most of the time, provided we see wood. But wood, as we have seen, is a superficial add-on. It is not a basic characteristic for differentiating plants. Flowers are, however, a good, but by no means infallible, guide in most cases. A pea flower is entirely different from a rose flower, and mostly on that basis we have a definitive signature that separates the pea family, the Fabaceae, from the rose family, the Rosaceae, to which apple trees and field spirea, a small bush, also belong. Similarly, after taking a look at one flower of a pea vine and one flower of a giant black locust tree almost everybody would consider them essentially identical, petal by petal,

A balsam fir seedling (right) and a ground pine.

in their whole elaborate construction. The difference between a pea plant and a black locust tree is not a basic design feature such as flowers. It is a trifle—wood. Wood also distinguished lycopods such as Lepidodendron trees from club mosses,* the collective name for ground pines and ground cedars.

Wood is a major evolutionary innovation of some plants for achieving tall stature. Stature with the aid of wood has been achieved again and again, in a great many of the world's most diverse plants. However, no plants are "born" or sprouted as trees. They only become trees after considerable addition of wood. At what size and after the addition of how much wood does a pin-size *Abies balsamifera* plant growing almost indistinguishable among the moss become a bush, and then a balsam fir tree? How big would a ground pine or a ground cedar have to be before we can call it a tree?

---

* The "club" from club mosses derives from their upright reproductive organs, attached to a long stem. These organs, resembling the male cones of pines and other coniferous trees, are technically called strobili. The club mosses, unlike conifers, produce reproductive spores, not seeds. These spores are microscopic, not large and edible like seeds. Predictably, the club mosses in any one area do not, like white pines, need to flower "sporadically" yet in synchrony with others of their kind to foil predators.

GETTING BY ON LESS

If one way be better than another, that you may be sure is Nature's way.

—Aristotle

Every year and through countless centuries, the leaves of the deciduous trees in my forest loosen their holds in early October. First one by one, then in droves, the golden sugar maple leaves, and the red maple leaves, colored lemon yellow, purple, vermilion, and orange, drift down and indiscriminately settle on the ground. The spent stems and leaves of the ferns, lilies, and other herbs add to the falling leaves. The paired leaves of the fingerling red maples turn crimson, orange, and yellow in October as do those of their giant parents. They then drop off, leaving matchstick trees with one small brown bud at the top. They become all but invisible among the ground pines, the balsam fir seedlings, and the tiny new white pines and red and white spruces.

About a month later, after the maples and other broad-leaved trees have shed their leaves, the sky usually darkens one day and the first fluffy snowflakes spiral through bare branches and settle on the recently fallen leaves. Shining and luminescent green against

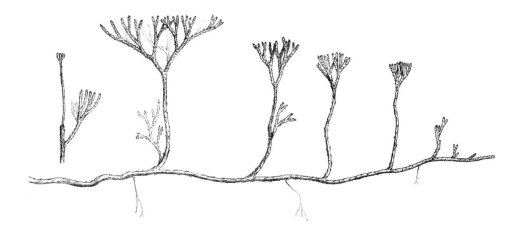

The ground cedar Lycopodium complenatum.

the new snow, the ground pines' and ground cedars' evergreen leaves remain unaltered, patient and enduring as before. The snow soon covering them eventually brings a halt to their photosynthesis, but it later protects them like a blanket from the winter's subzero air temperatures. In the spring the creeping lycopods come alive again. They send out shoots that creep along the ground, only sending up a small spike or sprig of a "tree" with a "club," a strobilus, on top that will release reproductive spores in the fall. The herbs, the trees, and the ground pines and cedars may grow together, yet all have different strategies for balancing their energy budgets upon which their survival depends.

Gaining height is advantageous in those plants that must grow taller than their neighbors to reach sufficient sunlight. Competition with neighboring trees that are also reaching up reduces lateral illumination and so the lower portion of a tall tree in a forest is relatively limbless. On the other hand, up top, where the tree has reached sunlight, it makes sense to put out leaves in a wide swath to collect as

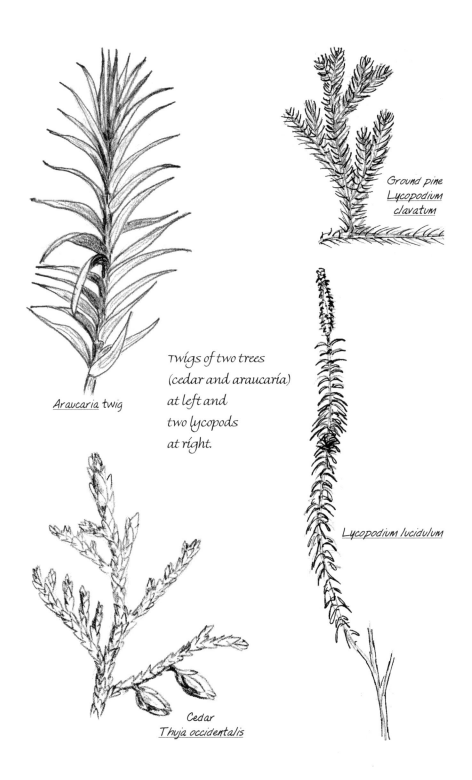

Ground pine
Lycopodium clavatum

Araucaria twig

Twigs of two trees
(cedar and araucaria)
at left and
two lycopods
at right.

Lycopodium lucidulum

Cedar
Thuja occidentalis

much light as possible. The lycopod club mosses of my forest, as already mentioned, don't even try, yet they still thrive. The mystery is, how do they do it?

The obvious and simple strategy of reaching high and wide has many costs, principally that of producing the massive and costly support structure. In northern climates (see pages 59–68) another serious danger of reaching high and wide is icing and snow loading and wind that can tear trees apart and/or can topple them unless they have a massive support structure. As a precaution against these dangers, as well as to reduce water loss, northern trees shed their leaves in the fall. The leaf shedding, in turn, is precisely what offers the lycopods that grow under them a brief opportunity to be illuminated by sunlight even as they are in danger of being buried by falling leaves. Since the light window is brief, and the lower temperatures in the fall and early spring slow down the photosynthesis factory, the lycopods are constrained to operate on a very low energy budget. They are not only exposed to small amounts of the sun's energy for a short time in the year, but also greatly hampered in their use of what little they do get. To see how they cope, let us again return to my forest in the fall for another look.

By the first week of October the colors of the trees are at their peak. Within a few days, as morning sun shines on the frosted maples, they are already shedding their leaves in droves. The slightest breeze knocks them off and they tumble down through the branches. The ground is soon covered and decorated like a multicolored Persian rug. At this point, the recently fallen broad leaves are still relatively moist and flat, but in another day or so they begin to dry and curl, then blow about in loudly rustling flocks. The leaves do not yet stay settled. The leaf cover shifts like the blanket of an uneasy sleeper.

As fall progresses, overnight frosts leave ice crystals on the morning grass. Sunny days give way to long slow drizzles and to occasional sleet and fog. Silver beads of water adhere to the waxy surfaces of fallen leaves, then are absorbed. The leaf blanket becomes increasingly moist and it settles softly down onto the ground, slowly enveloping fallen branches, tree trunks, and also small growing plants.

After the leaf blanket settles it smothers those plants that don't have mechanisms of coping with it. Spring flowers cope by poking through it, but mosses of all of the innumerable species can only grow where the leaf blanket slides off. Mosses grow on rocks and fallen trees that protrude above the ground, and on the lower portions of erect trees near the ground where there is still sufficient moisture. The lycopods escape by "running," hence another of their many common names, "running pines."

As a running pine spreads by growing forward on the ground, it creeps out from under the blanket then raises its growing tips upward at the edges. At intervals of several inches it sends one branch vertically upward and at the same time sinks a root downward. The falling leaves of the deciduous trees may eventually cover prostrate parts of the plant, but these loose fallen leaves do not balance themselves for long on those upright parts of the lycopod. The slightest breeze knocks them off. The covered *prostrate* parts of the plants' long bodies lose their chlorophyll and ultimately die, so that the live leading tip of the plant eventually trails a series of genetically identical clones. In aggregate they could be called one individual, and it could potentially be a huge one of almost any horizontal dimensions.

The lycopods can only run so fast and so far. Ultimately their dis-

Strobili

A running pine Lycopodium clavatum.

tribution in my forest reflects a precarious balancing act with their competitors. Almost none exists in either the unbroken evergreen or deciduous forests. In the evergreens there is no deep choking leaf blanket. However, there is obviously also no light window opening in early fall. In the deciduous forest, on the other hand, there is a brief light window in both fall and spring when the leaves are down, but there is always a thick smothering leaf blanket that is a threat despite the lycopods' running ability. Lycopods thrive in *mixed* evergreen-deciduous woods. By growing close to an evergreen tree, they do not become buried, yet the nearby deciduous trees allow sunlight to reach them laterally through bare branches in both fall and spring. (In temperate regions, the fall sunlight that illuminates them is low in the sky, as well as being from only a lateral direction.)

Lycopods gain their meager allotment of energy by being in the right place at the right time. As in any other organism, having enough energy is critical to survival. One strategy is to try to accumulate more, the other is to get by on less. Modern lycopods do both, but they have perfected the latter strategy.

TRIMMING THE DEADWOOD

We must heap up a great pile of doing, for a small diameter of being.

—Henry David Thoreau

The giants with whom the lycopods grow in my forest compete against each other in what biologists call an "arms race." This one is in wood production. Trees use wood as a scaffolding to hold their leaves above their neighbors, to get enough sun exposure to pay the price of growing more wood! But the ultimate goal is to procure a "profit margin" that will enable the tree to pay the costs of reproduction. In any forest the individual trees that manage to maintain a positive energy balance by reaching sufficiently high are in fact the exception. Competition prunes out the vast majority. Only one out of thousands or even millions of the offspring of any one tree will manage to garner enough resources to pay for sufficient growth to compete successfully and also to have energy left over for reproduction.

Trees are in a dilemma. They have to use their hard-won energy to produce wood (which is dead tissue) and often disposable leaves in order to grow, so they can obtain more energy. Usually this process

The competitive world of the forest floor (natural size).
The two lycopods (one young, one older) are taller than
seven deciduous trees (one to five years old) and
one spruce and one white pine tree.

results in a net loss, ultimately killing the tree. All gamble. Few win the energy jackpot of reaching sunlight that is a tree's equivalent to winning a lottery.

The genius of the lycopods is that they have evolved a way to avoid direct head-on competition against the giants, without producing wood. Nor do they need to make *disposable* leaves and branches.

We speak of deadwood, as if wood could be alive. In fact, all of the wood of a tree is biologically dead tissue. Wood consists of lignin-enriched strengthened cellulose cell walls, after the living material of the cell is removed. The only living part of a tree's limbs and trunk is a cell layer just beneath the bark.

Deadwood inevitably accumulates in any organization, tree or otherwise, where there is competition for resources. Unless consciously controlled, I suspect the call for deadwood would, as it does in trees, ultimately demand 99 percent of available resources. At a university, for example, one department commonly battles another for funds, to support the few "green leaves" in the organization, the professors who do research and teach. Inordinate amounts of time and effort are devoted to turf battles and administration in college, departmental, and committee meetings. There are memos that must be written, distributed, read, and responded to with other memos. There are so many meetings and memos that a whole class of people called administrators are hired. All leave a paper trail that, like the deadwood of a tree, is sterile.

Modern lycopods don't bother with expending all the energy it takes to produce a lot of deadwood. The economics of the ground pine resemble those of the logger using horses or oxen and a chain saw, as those of a great white pine tree resemble those of the logger with a million-dollar operation using feller-buncher, grappler-skidder and delimber-loader (all "modern" tools). The ground pine like the horse-logger has few overhead costs, so that it need not gather much energy to make a profit. As a result, it prospers yet has only a minor impact on others in the forest, and many individuals can therefore exist in a small space. But the white pine tree, like the logger with multiple expensive tools, can only exist by crowding out many competitors and it must amass huge volumes of wood merely to pay the overhead of staying alive. In the pine tree, the deadwood is not profit. It is the trunk and the limbs that serve to hold the crown to the light above the other trees. Both strategies persist, depending on circumstances. The big operators can persist *only* as long as there are big woods to harvest or potential big patches of free sunlight to reach, and/or as long as we create a state of nature through laws where we permit large-scale clearing. Without any of these conditions, they go extinct as surely as Lepidodendrons.

Hypothetical energy balance sheets can show how one strategy might persist in competition with another. But how and why have each of the different contestants developed its diverse strategies during the continually shifting landscape of evolutionary history? For the lycopods, major players on that landscape were, I suspect, the dinosaurs.

The large size of the lycopod and other land plants near the beginning of the Carboniferous age attests to an already long and intense escalation of competition to reach up for the light. By the time of the dinosaurs this contest to reach the sunlight had culminated in tall gymnosperms ("naked seed" plants), the descendants of a group that now includes our cedars, firs, pines, and redwoods.

The lush plant resources evolving on land under competition with one another provided vast fodder for herbivores who also exerted their influence on these plants. Taller plants survived better because they not only reached the light, but also were better able to escape the herbivores' hungry maws reaching up from below. The second and simultaneous escalating war between the herbivores and the plants likely contributed to gigantism (large size) in both. The larger the herbivore (such as brontosaurus, as an example) and the longer its neck, the higher it could reach up for food, and, in turn, the higher the plant had to place its leaves to escape.

After evolving the means to exploit this vast food supply, the first herbivores must have had a devastating effect. Eventually, they themselves became a target, a rich food source that spawned the evolution of carnivores.

Just as modern elephants do, the giant herbivorous dinosaurs probably trampled a wide swath of vegetation while feeding that disturbed the local habitat. This kind of disturbance could have provided opportunities and special niches for plants of small size. When ever-larger herbivores eventually could not afford to stoop for tiny browse, then *some* plants could escape by being too small to bother with. In the colonization of temporary clearings, the race would go to dispersers, to those who could be there first and who would reproduce before the sunlight was again cut off by the giant trees closing in from above.

The world of the clearing, be it caused by the feeding activities of a large dinosaur or by a storm's destruction, could have attracted a swarm of opportunists. The creeping lycopods were likely just one group of many of these opportunists.

The precise strategies to which different species appear to adhere with statistical probability are often thought to be "laws of nature," in analogy with our civil laws. That is, some assume that they are like god-given rules handed down from some higher authority that then must be obeyed. The reality of nature is precisely the opposite. What appear to be laws or rules are statistical probabilities that are the results of individuals that have responded appropriately to their own individual needs and contingencies. Most Lepidodendron trees competed in an arms race to reach the light above their fellows, as if that were the only law for survival that might therefore seem like a law of nature. And along came variants that disregarded this "law" by staying small, and they prospered.

The lycopods show us that in competition there is no "law of nature" stating that bigger is always better. Some black spruce trees in my bog grow only five or six feet tall, yet they still produce seed cones and can live for more than a century. On the other side of the continent, in the California redwood groves, a tree is a nonentity until it is about two hundred feet tall, and some have achieved phenomenal heights—up to three hundred feet. It is staggering to think of the thousands of years and thousands of tons of deadwood that a redwood tree must accumulate in order to just barely reach above the rest. The lycopods have triumphed in their own way. They have survived with few alterations more than four hundred million years because they operate with an energy economy involving only minute overhead costs. They have found a different way to avoid violating a fundamental law of survival: "Don't spend more than you have earned."

EVOLUTION OF SMALL VS. BIG TREES

Ecology is now teaching us to search in animal populations for analogies to our own problems. By learning how some small part of the biota ticks, we can guess how the whole mechanism ticks. The ability to perceive these deeper meanings, and to appraise them critically, is the woodcraft of the future.

—Aldo Leopold

Johann Wolfgang von Goethe said that it is so arranged that trees do not grow into the heavens. This is not good woodcraft. There is not one shred of evidence that nature arranges anything at all. Nature becomes arranged because the individuals it contains arrange. They accommodate and adapt. Besides, trees indeed grow into the "heavens," at least from the perspective of a ground cedar. Trees grow taller or shorter simply because of limits they experience that vary from one instance to the next. The limits imposed on them are largely those of their designs and of costs and available supplies. One of many designs that limits some trees from growing up to six hundred feet may well be their inability to transport sufficient amounts of water. In others the limitation must be that it takes too much time

An American elm tree,
as they used to look.

to grow that tall—time to produce enough wood before hurricanes and snow and ice storms level them. When the investments it takes to reach light are too great and too insecure because profits can be snatched away at any time, then the adaptive strategy (the one that on average persists) is to minimize investments. This was brought home to me after I saw all the dead and dying elm trees in my forest.

The American white elm, *Ulmus americana*, was one of the largest and most distinctive trees in eastern North America. It grew to a height of 125 feet and a diameter of 7 feet. Its tall straight trunks and umbrella crowns with drooping branches made it a favorite shade tree of town and city all across the land. Elms dominated the yards of the two farms where I grew up, and also an outdoor chapel.

William Cullen Bryant once said that the groves were God's first temples (in "A Forest Hymn"), and the Reverend George Walter Hinckley, who founded the Good Will School in Hinckley, Maine, must have listened. "GW," as we called him at Good Will, where I spent six of my boyhood years, built an outdoor chapel by planting elm trees into the configuration of a broad cross like four squares abutting a central square. The whole array was framed by spruce trees. I remember these elms grown together at the top to form a green "ceiling" more beautiful than that of any chapel, because it contained live singing birds. It shaded the "floor" and a pale green light filtered down. It was a safe and protected place and reminded me of a hollow under a patch of dense, leafy brush that I used to hide under when I was much younger. This elm chapel could inspire a deeper appreciation of life than any built of plaster, stone, and glass.

The elm chapel was built, I presume, to stand for hundreds of years. But it is gone now. The "Dutch" elm disease (probably a native of Asia that was first found in the Netherlands and then spread to America in 1930) destroyed the chapel. Tiny brown beetles burrowed into the bark. They carried with them the spores of the fungus *Ophiostoma ulmi* that spread in the trees' system of tubes that, like arteries, carry fluids and nourishment. Eventually the fungus plugged the tubes. The first symptoms that the trees showed were a few yellowing leaves on some twigs, then more, and more. . . . Eventually, whole twigs died and bacteria broke down the cellulose of the wood. Twigs fell. Long-horned beetles were attracted to the smell of the dying, defenseless tree. They laid their eggs on it, and their larvae burrowed into the bark, then chewed into the wood. They too carried bacteria and fungal spores deep into the tree. When all of the branches had broken off, only tall bare trunks of the trees remained. Pileated woodpeckers hammered oblong pits into the sides of these columns, excavating yellow juicy long-horned beetle grubs. The fungi grew ever-more vigorously now that water had access to the trees' interior. Their slender tendrils grew up and down the trunk, consuming even more of the wood and pushing out spore-dispersing fruiting bodies, the shelflike mushrooms colored in pastel reds, browns, and yellows.

By the 1990s, I could find only one of the large elms remaining near my forest on Adams Hill. It stood by a farmstead along Route 156, between Wilton and Weld. Then, in the summer of 1992 the telltale dying leaves appeared on its drooping branches in July. I took the tree's photograph. By the next year the farmer on whose land it grew had cut it down for firewood and placed a large flowerpot of

*Dead elm tree trunks with peeling bark and flicker nest holes on top,
and with the typically oblong pileated woodpecker excavations below.*

geraniums on its stump. This tree was well over a hundred feet tall, and the diameter of its stump measured more than six feet.

The American chestnut had just been swept from the American continent's forests by another Asian fungus (first detected in 1904 in the New York Zoological Park). Nothing, it seemed, would now stop the extinction of yet another dominant tree species by the inadvertent introduction of an alien organism. The pathogens may have come in only a microscopic cell. But one organism, even one cell, packages enough information to wipe out the entire population of a tree species over the entire continent, despite all human efforts to control it. Annually, millions of acres in the Northeast are defoliated by the European gypsy moth caterpillars. A long-horned beetle from Asia has just been discovered killing maples in New York. An introduced fungus is damaging beech trees. Another kills birch. There are now also frightening reports of new epidemics starting up in hemlocks and dogwood, all from introduced biological agents that have invaded through the Trojan horses of imported logs, lumber, or seedlings. I wonder what may come next.

If a population is dense, pathogens can spread easily within it, moving like wildfire in an unending sea of dry grass. The only escape may then be isolation. My forest is now the isolated haven of four American chestnut trees that

What elm trees look like now (most easily identified by sideways bent terminal bud on each twig).

were raised disease-free on sterile soil before being planted here, near their former haunts. Here they have so far escaped the blight. They are growing vigorously and this year they flowered.

Might the elms somehow also be resurrected? At first, it seemed to me that they had indeed already gone extinct. I no longer saw the large and distinctive forms I had grown up with and had assumed were an inviolate part of the landscape. However, one day as I was looking at the forest without expectation I was surprised to discover that elms were still quite common. I only needed to search differently. The elms of the nineties were not in the stately form of forest and town I'd been used to. Now, the wild elms are growing mostly along roadsides. And they are virtually all juveniles reaching straight up with their branches, not achieving those majestic contours of adult trees. Was the fungus selectively killing only the larger trees, or were the younger trees still alive because the disease had not yet reached them?

In order to see more clearly what the pattern might be, I counted the elm trees along about four miles of country road near my house in Vermont. My tally was a total of 862 elm trees or their dead remains. All of the largest trees had been dead for one or two decades and were almost fully decayed. Of the 540 *live* trees, only 47 had a diameter greater than six inches. That is, virtually all the big trees had died and the living trees were small in size; the probability of still being alive when the tree was three inches in diameter was 97 percent, but when a tree had grown to six inches its chances of still being alive had dropped to only 60 percent. When a tree reached a one-foot diameter, survivorship had dropped precipitously to 7 percent. Given the low odds of surviving until reaching a one-foot diameter, it would make sense for the elms to reproduce as young as possible. Clearly, if a tree was going to leave offspring, it needed to do so when it was still small, not when it was mature. But can young elms reproduce?

The American elm blooms in April. It is one of the earliest trees to bloom and its small wind-dispersed fruit are mature long before the leaves unfurl. While gathering data on trees' size and health I also recorded reproduction (flowering). One hundred and twenty-five of the 540 live trees had blossoms. Surprisingly, blooming had occurred in very small trees (4.4 percent of the trees up to three inches in diameter, and 47 percent of those of three to six inches in diameter). Those few trees that were over six inches in diameter were *all* in bloom.

If a tree has *assurance* that it can grow to a large size then there is an advantage for it to invest its energies in growth because large trees have access to the most sunlight and can then afford a large reproductive effort. If it is inevitable that the tree will be killed before it reaches a half foot in diameter, then only those who have a tendency to reproduce early, while they have the chance, will pass on their genes to future generations. On the Vermont roadside, the advantage of waiting to grow to a ripe old age was nearly zero. Perhaps the best strategy for elms now may be to reproduce at any size, immediately upon becoming infected. In that way at least one crop—perhaps the largest possible in a lifetime—would be achieved. In a hopeful scenario such as the one I have imagined here, given enough time, a parasite (the fungus) could evolve to become a symbiont: Theoretically a tree could evolve to *require* the fungus as a signal to tell it to bloom.

In the forest symbiotic relationships are legion from the bottom up. For example, the cell of each leaf contains fifty to one hundred chloroplasts, the microscopic green structures that are the plant's solar energy collectors. Now symbionts of plants, they were once algal parasites. Mitochondria (the even tinier bodies inside an organisms' cells that convert the energy in chemical bonds to other forms that can be metabolized by the organism) are evolutionarily tamed symbiotic bacteria that were once also parasitic invaders. Algae infecting fungi became a new association called lichens. Lichens absorb nitrogen from the atmosphere, making it available to other organisms. Fungi infecting some tree roots ended up in associations called mycorrhizae that help the roots to absorb nutrients from the soil. Blue jays and squirrels doing their level best to eat every acorn they find became uphill planters. By dispersing the trees' seeds, they have become symbionts. All of these associations, and many thousands more, are combined in unimaginably complex ways to produce the ecosystem we call forest. It is perhaps the most complex symbiotic relationship on this earth.

The trees now coping with Dutch elm disease nudged me into thinking: Why did some species become so large? Few of us would dispute that large size in trees has resulted from competition to reach the light where small size and shading meant failure. What is curious is that gigantism of trees is a phenomenon of *western* America, with maximum heights of 365 feet in redwood, 329 feet in Douglas fir, 315 feet in Sitka spruce, 310 feet in giant sequoia, 261 feet in western hemlock, and 158 feet in big-leaf maple. Their ages are commensurate with their size. Here in the East, all the tree species are midgets compared to those out West. Few reach the height of the western trees, and none grows really old. Why? Perhaps the current predicament of the elms is instructive. There is only one compelling reason for a tree to reproduce when young and small, and that is when the chances of its growing old are slim. When investment for the future is insecure, then all survival mechanisms concentrate on the immediate present only. In my forest there are not, and never were, trees older than about three or at most four hundred years. A two-hundred-year-old tree is an anomaly. Why aren't there trees that simply keep on growing for another hundred, or thousand, years, to tower 300 feet up into the air? Are they likely to be blown over by hurricanes?

The northern forests have historically been swept by periodic hurricanes that have leveled them. In my forest, strong winds take hundreds of the larger trees every year. Ice and snow storms break down limbs and topple those trees that reach too high, making it costly to be so tall. Here in the East, any tree that waits to reproduce until it is hundreds of years old will have a slim chance of reproducing at all, and having left offspring there is reduced selection for it to live longer (see pages 164–165).

My white pines are near fifty years old and they have not yet flowered. For now, they are locked into a do-or-die competition to reach above one another into the light. Those who make it might flower for two hundred years. Those who don't might die next year. Producing a big cone crop *now* could be suicidal, because that would slow a tree's growth, permitting the neighbors to gain a lethal advantage. They all gamble for a long life, one that a tree less sturdy than a pine cannot chance. Investment for the future is determined by security in the present and promise of a future.

CONSTRUCTION FOR STRENGTH

When I see birches bend to left and right
Across the lines of straighter darker trees,
I like to think some boy's been swinging them.
But swinging doesn't bend them down to stay
As ice storms do. . . . they are bowed
So low for long, they never right themselves:
You may see their trunks arching in the woods
years afterward. . . .

          —Robert Frost

The storm on January 22, 1995, was by no means as bad as it gets, but I found it particularly annoying. During the cold rainy night my research aviary collapsed under the weight of thick ice that had collected on the wire screening. In the morning, ice hung like crystal from the trees—a pretty sight, except that several limbs of my favorite white birch next to the cabin were dangling limply, having snapped under their loads of ice. Six-inch-thick gray birch trees were bent over double, their tops touching the ground.

Surprisingly, most of my trees remained intact. Unlike my

aviary, a temporary construction of wood and wire, the forest still stood. Nothing attracts less attention than complete success. However, thanks to the aviary and some broken limbs, as I now surveyed the icy trees, I began to see the survivors as examples of superior design shaped by natural selection; any inferior tree models have long since been obliterated from this landscape by the climate.

Bridges, parking garages, and skyscrapers rarely collapse even under extraordinary circumstances. That is because engineers have extensively tested the beams, bolts, and trusses that go into their construction. They have searched for every conceivable weakness and subjected the materials to stringent tests. Construction can then be specified within tolerances that not only meet but exceed anticipated stresses.

Trees have also been tested throughout history and they are constructed to stand up to specific stresses, but *not* to exceed them. They have been selected to withstand the power of the wind, and in northern climates, to withstand being loaded with ice and snow. Unlike bridges and skyscrapers, they have a smaller margin of safety because, as in all construction, safety is bought at a price. In the fierce competition among trees, compromises have to be made. Trees cannot afford to insure themselves against all possible stresses. Their optimal solution is to pay for "just enough" to ensure survival given exposure to "average" stresses. So trees are constructed to withstand not all, but most stresses. As a first requirement for reproduction a tree must remain upright, which it usually manages to do. But just what are the limits of tree construction? What makes them break or fall?

Here in Maine, one of the recurrent physical stresses that trees face are ice storms. The conditions of this particular ice storm on Adams Hill were just severe enough to point out the weaker aspects of tree design and to highlight the strengths. I recalled an ice storm about ten years earlier that had left the ground in the mature hardwood forests a huge tangle of broken limbs. Gray birches had been broken or had their tips bent all the way to the ground. The white birches, with stronger trunks, had great limbs torn off. I don't remember ever seeing a limb of a fir or spruce tree broken off; moreover, those trees always remain erect. This latest storm affected only the white and gray birches; the red and sugar maples, American ashes, beeches, and yellow birches—all fairly young trees—were not damaged.

Several days after the storm, when I wandered in a relatively mature forest nearby at Perry Hill, I discovered much more destruction. Here some mature, seed-bearing oak trees, beeches, maples, white birches, and thick-trunked yellow birches had toppled over, and the ground was littered with fresh branches. Thick limbs of large pines had snapped off. So, while most of the trees were clearly "working" just fine in this particular storm, the gray birches near my cabin and the mature hardwoods at Perry Hill reminded me not to take design for granted.

I wondered what principles were at work here. The first of many variables was the amount of ice the limbs had collected. A look out the window suggested that the birches had collected a lot more ice than the other hardwoods, so I set out to see if this was the case. With bush cutters, I carefully snipped off five three- to four-foot-long branches of five different species of young hardwood trees, including white birch. Careful not to knock off any ice, I weighed all twenty-five ice-laden branches with a spring balance—the kind fishermen use to weigh their catch —then spread them all out on the floor of the cabin. By late afternoon, the ice had melted, the cabin floor was awash, and I weighed them again.

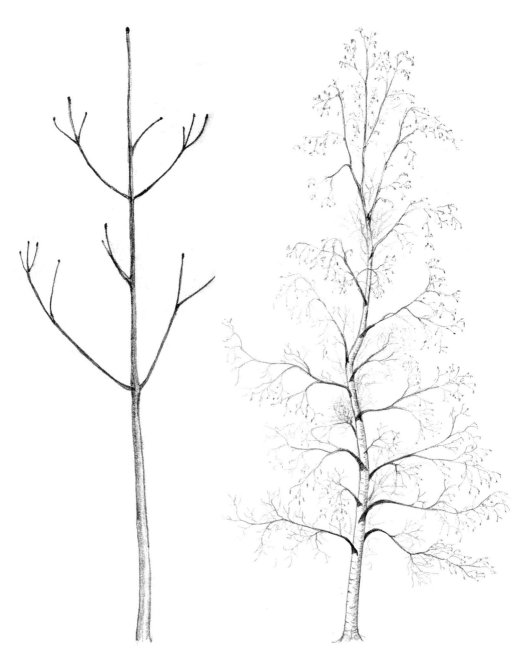

A young ash tree has few twigs and a candelabra shape that minimizes ice-loading on the branches.

Birch trees (young gray birch) have many twiglets that bend down when loaded with ice, which facilitates even more ice-loading.

As I had suspected, per unit of twig weight, birch collected much more ice than other tree species, dramatically so. On average, a birch twig carried a load of ice eight times its own weight. American ash had the lowest ratio of ice weight to branch weight, while sugar maple, red maple, and apple twigs had an intermediate ratio.

Young ash trees have the fewest twigs, which may explain, in part, why they collect the least ice. Birches, which collect more ice, have bushier and more numerous twigs, so per weight, they have more surface area for ice to adhere to. Moreover, when birch twigs are partly ice-laden, they tend to droop outward, away from the stem of the tree. Surface tension holds water droplets on the twigs longer, and the slow flow down the twig allows a thin film of ice to form. In contrast, the relatively stiff, horizontal twigs of apple trees, for example, generally hold little ice. The water simply drips off.

A vertical rod has only its tip exposed to rain. Laid horizontally, its whole side is exposed. This principle is important in the architecture of the tree. Ash, maple, and poplar trees (especially younger ones) have upward-pointing twigs, and this arrangement of twigs has two effects with respect to ice. First, the more a twig points straight up, the less surface it presents to falling rain (that freezes to ice on the twig), and the less ice will collect on the twig surface. Furthermore, the branches form rough Vs relative to the trunk, so rain that does land on the twigs runs *inward* toward the trunk (rather than *outward* as on the gray birch) and freezes where it can do the least harm. When the ice is at the tip of the branches, the force of gravity is magnified by leverage. Close to the trunk, the force may become sufficient to break the limb. The closer the ice forms to the trunk, the less leverage it can exert. Indeed, due to the V architecture of the trees, the ice on most young maples, ashes, and poplars was concentrated on their trunks. Possibly this ice even helped to stiffen them. If it were collected on the outside of branches (as in gray birches) then it would weaken them instead. In this particular ice storm, none of the candelabrum-shaped trees showed any sign of being burdened by an ice load, nor did they accumulate much snow later on. Mature trees generally tend to have large spreading crowns, and as one might expect, they lose many limbs to ice storms while the thinner and more upward-pointed trees go unscathed.

Most of the surface area of the trees in my forest is, however, not in their twigs, branches, and stems. It is in their leaves. To shed leaves would be to remove the major site for dangerous ice-loading. I wondered if the deciduous habit of the oaks, beeches, basswoods, and other broad-leaved "hardwoods" is an adaptation to circumvent this danger.

It is easy to speculate. However, as with many other evolutionary questions, this one is not easy to answer. First, trees shed their leaves so the effect of *not* shedding them on the same trees is hard to measure! Second, several selective pressures may act similarly and simultaneously, so that it is difficult to tease apart their individual contribution to evolution when they've acted together for millions of years.

The best approach to evolutionary questions is usually a combination of observations, experiments, and comparative studies. As it turned out, the "experiment" of what happens to deciduous trees under ice loading when they *do* have leaves was later done for me. And the results were dramatic. It happened on the night of May 30, 1996, near my home in Vermont, where many deciduous trees had already put on their leaves about three weeks earlier. (They were two weeks later in 1997.)

Early that night I heard rain pummeling the forest of lush green foliage. Temperatures were near freezing, and kept dropping. When

An elm twig, with new leaves, pulled down by a load of fresh snow on May 31, 1996.

I awoke and looked outside at 3:30 A.M. I saw a changed world. All was white. The green leafy trees were all laden with snow and bending over in all directions. The nonwoody green vegetation—the lilies, and other various wood flowers without woody support—were already flattened. The slender young maple, oak, ash, and birch trees were all bent over double, touching the ground. Only the conifers—firs, spruces, and pines—stayed erect. The big-stemmed deciduous trees had branches broken off. Later that morning the roads were strewn with great torn-off limbs and with toppled poplars and maple trees. Crews were working with chain saws to clear the debris so that cars could get by.

All of this was the result of only three inches of snow. It was a dramatic illustration of what happens when deciduous trees get caught in a relatively minor snowstorm with their leaves on. I had the impression that a great many of the remaining limbs of the deciduous trees were near the breaking point. Had temperatures been only a degree or so lower, then much more of the rain would have turned to snow and there would then have been vastly more destruction.

The unplanned "experiment," or perhaps the planned observation, depending on one's viewpoint, left no doubt that it is costly for a tree to keep leaves in a northern winter. Leaves are, of course, also costly to lose, because it takes time and resources to grow a new set each spring. The farther north one goes, the shorter the time that is available in the summer not only to grow new leaves but also to take sufficient advantage of them after having made them.

Leaves have evolved to stay on in the winter, under situations where snow and ice-loading does little damage. In my forest and the bog near it, the ground-hugging lycopods, Labrador tea, bog rosemary, wintergreen, as well as spruces, firs, and pines, all keep their leaves for at least one winter. (White pines keep them for two winters, and balsam firs and red spruces keep them for at least five and up to ten years.) On the other hand, winterberry (a holly) and mountain holly both shed their leaves each fall, while other more southerly holly trees and bushes keep theirs through the winter. The red oaks in my forest lose their leaves, while many southern oaks don't. My red oaks hang on to their leaves for much longer than all the other deciduous trees as if they had not yet perfected their technique of shedding.

The evergreen trees appear to be special. They have managed to keep their leaves all winter *and* largely to avoid toppling under snow and ice-loading. How can they possibly do it?

The answer is in tree design. The main feature that protects spruce and fir, and to some extent pine, is the very feature that we cherish in them as Christmas trees: their conical shape that is achieved by the trees' growth pattern. Each spring, the trees sprout a single, straight shoot from the top. At the same time, a whorl of three to six branches grows off horizontally. The already existing lower branches stretch even farther to the sides. The number of umbrellalike tiers of branches, with the largest on the bottom, corresponds to the age of the tree.

Spruce or fir tree unloaded vs. loaded with snow.

Conifers can keep their leaves in winter because this shape allows them to shed their load of ice and snow like an umbrella. As snow and ice begin to collect on the leaves and twigs, the horizontal limbs begin to droop. Each tier droops until it touches the tier below, which provides support. The "end loading" of the conifer branches is not a liability but an asset. The upper whorls of branches press down and are supported by the lower branches, resulting in a stable cone, or tepeelike structure. This design is possible because the tree has only *one* trunk growing straight up, and lateral branches taking a sub-servient position off to the sides.

Folded against the trunk by the weight of ice and snow, conifer boughs act like a collapsing umbrella; they then intercept less pre-cipitation. In spruces, the same principle is applied on a small scale to the individual branches. This is a most desirable trait for a tree in Maine. The more load the tree takes on, the more snow slides off, just as it slides off my steep cabin roof or off the sides of a tepee.

Differences in wood that affect the limbs' strength are another important variable. As an inveterate tree climber since my child-hood, I feel qualified to comment on the strength of tree limbs. I've hung by one hand from slender, inch-thick live spruce limbs or jumped up and down on them in the top of spruce trees. There was no danger. However, jumping on a white pine limb of the same thick-ness would be treacherous. Every winter the white pines in my for-est lose many limbs.

White pine is not totally evergreen. It sheds *half* of its leaves each fall. Apparently what it lacks in limb strength, it usually makes up in limb thickness. The limbs of tamarack (larch) that grow in a dense stand down by my bog are even more brittle than those of pine, and thinner as well, but this conifer sheds *all* its leaves in the fall. I have no doubt that if tamarack limbs could accumulate as much ice and snow as spruce and fir limbs do, they would regularly be snapped off the tree. This rarely happens. Being deciduous, tamaracks can appar-ently afford to have weaker limbs. Thus, by a combination of stratagems, each tree is sufficient to withstand the challenges it com-monly faces, but there is no point for it to do more than is usually necessary.

WOOD

*The greater part of the phenomena of Nature are . . . concealed
from us all our lives. There is just as much beauty visible to us
in the landscape as we are prepared to appreciate, not a grain
more. . . . A man sees only what concerns him.*
—Henry David Thoreau

The flexibility of a birch, the tallness of a pine, the toughness of an
oak, all reside in that marvelous substance we call wood. It makes a
tree a tree and it has virtually limitless uses. Yet, when I walk in my
forest of wood, I hardly see *what* this substance is at all. To really see
wood, I need a microscope, a chemistry lab, and a good reference
source.

Such wood as that from a balsam fir tree is a system of narrow
tubes composed of many hollow cells called tracheids placed end to
end. The tracheid cell walls are composed of cellulose (a glucose
polymer) and these cells are thin-walled in the spring when the tree
grows rapidly. They are thick-walled in the fall when it grows slowly.
It is this difference in cell-wall growth that results in the annual
growth rings that we see in temperate and northern trees. These
growth rings are laid down by a single layer of dividing cells just

under the bark. The cells laid down toward the inside after each division become wood; those to the outside become bark.

The wood tracheids are very long and narrow cells. An average pine tracheid is about 7 millimeters (mm) long and 0.2 mm wide, but the dimensions vary tremendously between different tree species, from 0.4 mm to 18 mm in length and 0.1 mm to 2.0 mm in width. The tracheids' nuclei and cytoplasm degenerate when the cells reach maturity, leaving only cell walls as dead, hollow tubes or pipes. These are aligned end to end, forming continuous tubes. Fluids flow through these tubes between roots and crowns, but how that is accomplished seems a miracle. How, for example, is water lifted some 200 or more feet to the tops of redwood trees? The world record is probably 435 feet, the height of an Australia mountain ash, the tallest tree ever recorded. It must require energy to somehow lift water to such heights. We now know, however, that more simply and perhaps even more remarkably trees do not need to expend energy to transport water-dissolved nutrients from the soil and up into the leaves.

The water flow is actively regulated at the very top, by the opening and closing of the stomates or pores in the leaves. When the stomates are open water evaporates from the leaf. Each water molecule is part of a continuous unbroken column within the tracheid tubes, reaching all the way to the roots. The water molecules adhere to one another. They are anchored to other water molecules and also to the cellulose molecules of the tracheid walls. Each time a water molecule evaporates from the top of the chain through a leaf stomate, it pulls the rest of the chain behind it. At the bottom of the chain, at the very root tip, a molecule of water is pulled out of the soil to fill the "gap" in the chain. Evaporation from the leaf surface drives movement of the water column. In a sense, the tree forms a continuous system with the soil in which it grows and the air that it breathes. All is beautifully engineered to take advantage of the simple physical properties of adhesion and evaporation. It is possible because of the remarkable macroanatomy of the wood.

The dead tracheid cell walls, which end to end form the water pipes, are the wood "fibers" used to make paper. One cell, one fiber. The fibers themselves are composed of long cellulose molecules coiled in a helix around a hollow axis. If arranged end to end, about 2,000 cellulose chains, each one of them composed of about 1,000 glucose or sugar units strung together, would equal about one tracheid length. There are about 2 billion (2,000,000,000) cellulose chains in one tracheid or wood fiber. Each tracheid cell thus contains the equivalent of 2,000,000,000,000 sugar molecules, which are the primary product of photosynthesis. Most of a tracheid's growth occurs in about thirty days, thus one tracheid cell will add an average of 2×10^{12} glucose units/(2,592,000 seconds/30 days) = 771,600 glucose units per second per tracheid cell for the entire thirty days. Glucose is a sugar composed of six carbons, requiring six carbon dioxide molecules to make, hence one growing tracheid cell takes up the equivalent of $771,600 \times 6 = 4.6$ million carbon dioxide molecules per second.

These staggering numbers suggest a story. The carbon dioxide is taken up from the air, masses of which sweep across continents in a matter of days. Since these air masses mix together as in a giant blender, it stands to reason that those 4.6 million carbon dioxide molecules taken up by just one tracheid cell (say in a twig of a maple seedling next to my cabin) in just one second could have come from a decaying log in the Amazon, a car on a Los Angeles freeway, a coal-burning power plant in Utah, a hornbill in Indonesia, and a baboon in Tanzania. If we could put a pinhead-size red dot on a map of the world to indicate the source of each of those 4.6 million molecules that were produced just in the last week by one growing tracheid cell, then the whole map from pole to pole would be colored solid red. Only some regions would be colored less deeply red than others. Conversely, if we colored the map with blue dots for the fate of each of the oxygen molecules that the tracheid produces at the same time, then in a week or less, the earth map would be a nearly solid blue. Each wood cell of every tree in my forest is in a give-and-take with the rest of the world.

Biochemically wood is a "composite" structure—one that is made of two separate chemical substances, cellulose and lignin. In simplest terms, the cellulose makes it flexible and the lignin (the "glue" that binds the cells together) makes it hard. The two together give wood its wonderful combination of qualities: strength and flexibility.

The cellulose molecules of a tracheid bind water lightly by a process called hydrogen bonding: the positive charges of the hydrogen are being attracted to the negative charges of other molecules. "Hydrated" cellulose, that with water lightly held to it, is much more flexible than dehydrated cellulose. Hence a living tree can bend in the wind without breaking. The lignin of wood is a polyphenol, that is, many phenols bound together. Lignin also binds lightly to the cellulose by hydrogen bonding. When the bonding sites available in cellulose are not binding water or lignin, they bind instead to other cellulose molecules. This bonding has important consequences. For example, when we wet newspapers (a dried water suspension of wood fibers from which the lignin has been removed), the hydrogen bonds between cellulose molecules are now taken up by water, so that the paper disintegrates.

Like paper, cotton, linen, rayon, and cellophane are also nearly pure cellulose, that is, they are long chains of sugar molecules attached end to end. Commercially prepared cellulose, unlike most paper, is transparent because the normally helically coiled cellulose molecules found in the wood tracheids have been uncoiled so that they are flat and no longer contain air bubbles that cause the substance to be reflective rather than transparent. Cellulose from trees can also be made into explosives, by attaching nitrate molecules to it (cellulose nitrate).

The width and length of the wood fibers and the amount of lignin vary among different trees, which results in vastly different properties of strength, hardness, and flexibility. The paper industries generally prefer wood with a minimum of lignin, because it must be extracted to make paper, an expensive process. In addition, ideal paper wood has long tracheids, for greater strength. The wood of the yew, which was long considered the ultimate weapon of war because it made the best English longbows, is flexible and strong presumably because it has relatively little lignin and long fibers (tracheids). The wood of the black locust tree has short fibers and much lignin so that it cannot be compressed. It is stiff and can be pounded like nails. It was used to make stout hard pegs for bolting the planks of wooden ships together.

I have two such locust trees next to the cabin in my forest. They came from a farm in northern Maine, near Wytopitlock, where my late friend Phil Potter boarded as a young man when he worked in the woods squaring cedar railroad ties with a broadax all day, all winter. He showed me the farm on a fishing trip we took to that country. We dug up root sprouts and took them home in the trunk of his Plymouth Duster, planting them on the lawn of his house. My two locust trees are sprouts from those we dug up together. I've planted them not to use as a substitute for nails, but because they connect me to a history that I don't want to forget.

TREES AS INDIVIDUALS

He who plants a tree
plants a hope.

—Lucy Larcom

My son Stuart and I picked a red oak tree, because oaks live long and grow large. The tree we picked in August of 1994 had probably been planted originally by a blue jay because it was growing in the shade of white pines a half mile from any mature potential acorn-bearing parent. We cut into the soil around it with a spade, severing the major lateral roots about a foot and a half away from the two-inch trunk. We needed to retain enough roots to give the tree a new start in the new hole we had dug for it in the open space to the southwest of the cabin. We pulled off half the leaves, so that the tree would not lose too much water, given that most of its root system needed to be regenerated before adequate water delivery could resume. We also thoroughly soaked the earth we planted it in, then covered the fresh soil around the relocated tree with a mulch of dead leaves, bark, and sawdust.

The tree scarcely grew at all the year after we planted it. It was reestablishing roots and recouping its losses from the stress of trans-

The young red oak we planted by the cabin, showing the characteristic branch pattern on March 5, 1996.

planting the previous August. Although it leafed out, several of its branches died. The second summer it grew several new shoots from near the bottom of its trunk. I sketched the tree, noting the branches it had lost and the new ones that looked promising.

The oak sapling is now on its way to becoming a tree. I will not trim any of its branches, letting natural forces shape its character. I look forward to its progress for as long as I live, and perhaps some-day to see a big old oak laden with history and acorns!

Most of the trees in my forest are much too young and perhaps too anonymous for me to notice their individuality. However, now I will know this oak. History was deliberately created. Since we shared in the act of planting it, my son and I had made a bond with this tree, and our relationship to it is now special.

We remember the mountains we climbed and the ponds we swam in on warm sunny afternoons when we were young. We remember the bend in the river where we caught a trout. We see permanence in the hills we get to know. They do not visibly change, but we can grow up with trees and see them change with us. Since they can outlive us, they connect us to the future and to the places where they grow.

I recently visited a forest where I had last been as a child. Much of that forest had become unrecognizable to me. But walking down the sandy dirt road in those woods my mind's eye suddenly and startlingly recalled an old beech tree with a double trunk and a pecu-liar twist. I saw myself at nine years old collecting nuts under that beech with my sister and mother. Further along, around another bend, there it was! I recognized it by its peculiar trunk. That trunk was the tree's unique signature. Recognizing it was like recognizing an old friend from the distant past, as opposed to seeing just another face in the crowd.

Like us, each individual tree is genetically distinct in ways we do

An old (approximately 250 years) red oak tree in the forest. Branches have broken off at various ages, creating animal den sites.

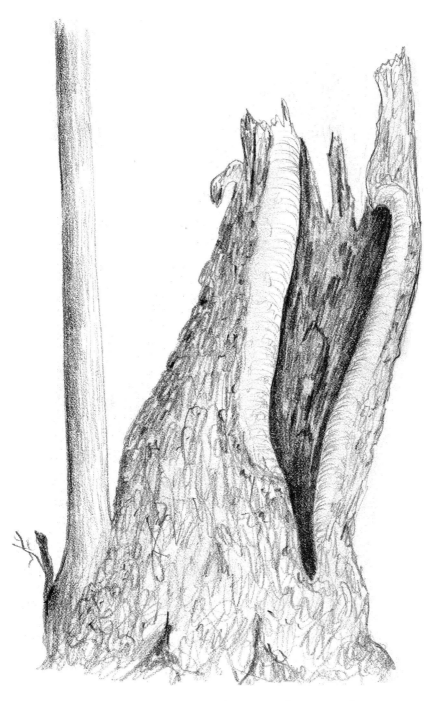

A large Amelanchier or Juneberry tree that grew to have a
circumference of forty inches. The old tree was hollow, then broke off.
A new shoot grew up from its stump (left).

not normally see. However, much more noticeably, a tree is shaped by its experiences. Like us, the longer it lives the more different and independent it becomes from other trees. A record of what has happened to it becomes inscribed on its body. Wrinkles in the bark reflect age. Branch thickness indicate the loads it has carried and the directions from which it has experienced light. Scars remain where overloaded branches broke off and where moose and caterpillars attacked. The width of growth rings may reflect droughts, crowding, temperature, and prolonged tilt or lean. The tree's general shape and pattern of branch growth result from the competition it has experienced in its struggle for light. There is a local explanation for the heft and warp of every twiglet.

The white or paper birches common all over my forest are the trees that strike me most with their individually distinctive form. Their brilliant white trunks and large limbs stand out starkly against the dark background of the forest in fall and early spring. I chose four birches and sketched their major outlines. They were aged forty to fifty years, but they have had very different histories. Two of these trees grew up in the open. They had easy access to sunlight and so had spread wide, but this growth pattern had made them vulnerable to ice storms. The loads of ice had ripped off branches, and the scars where the branches had been were still visible. Two others had grown up under intense competition. They grew straight up, striving to reach direct sunlight. Their major branches were at acute angles to the trunk. This had helped to protect them from ice loads and these trees had few scars from broken-off branches.

Every birch, every old oak tree, is different from every other tree that ever lived and that ever will be. Each tree has branches and the branches have twigs and the twigs have leaves and buds and fruit. The precise number, thickness, configuration, and angle of branches

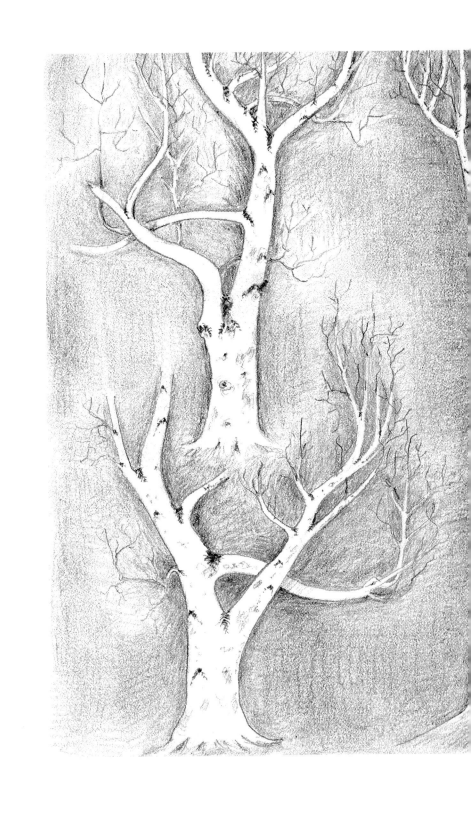

Four white birch trees with individually distinct branching patterns.

and twiglets are not predetermined. Nor are the numbers of leaves, flowers, or fruit predestined. Nobody knows beforehand which branch may break, and when, and how each broken branch may induce compensation in growth to realign the tree's shape. It is by its experiences that an old mature tree becomes, like a human face, distinctive. Even though all trees are individually unique, we seldom have reason to notice individuality in those that do not intersect our personal history. Such details are scientifically tedious and meaningless. Nevertheless, the ability to discern differences (and similarities) as they relate to species are profoundly important to biology. These differences also relate to history, but to the long evolutionary history of life, where random chance plays just as much a role in determining a species's characteristics as personal history determines a tree's or our own.

VINES

I like trees because they seem more designed to the way they have to live than other things do.
—Willa Cather

Trees, like ants on a scent trail, act as though they "know" precisely where they are going from their earliest youth, and they pursue that goal relentlessly. Their goal is to reach the light by whatever means. For many plants, the strategy of choice is to take advantage of the investment of others by using the others' costly scaffolding of wood to reach light. There are those who dispense with this investment, or minimize it, and reach just as high as trees. They are the climbers. We call them vines. They are especially numerous in the tropical rain forest.

In anthropomorphic terms, the "meanest" of these hustlers is the strangler fig. It starts out like a vine that appropriates scaffolding, rising fast and high to reach the light in a twinkling of the time that its host needed to reach the same height. Then, having achieved the same prominence as its host, it suddenly bulks out while basking in the sunlight and shading its host. It becomes a killer tree. There are some maple trees in my pine grove that start out vinelike as well.

A young sugar maple tree climbing vinelike into the crown of a white pine tree.

The pine grove just to the west of my cabin sprang up only in the past fifty years, from a cow pasture. The pines colonizing the pasture originally had limbs down to the ground. A sugar maple seedling growing next to one of these pines happened to lean onto the branches of a pine. Since the maple was then physically supported, it could afford to invest less wood into making a thick stem for support. Instead, it used the meager energy supplies that it had to grow upward, to shoot for the light. Several such maple "vines" that I found had diameters of only two to three inches but they had grown to twenty or thirty feet tall. They leaned crookedly against successive pine branches. White pine trees eventually grow taller than any sugar maples, so the prospects for these thin maples were not good. However, this stand of white pines was infected with blister rust, and many trees would gradually die because of this disease. Then the dead pines would continue to stand upright for another ten or more years. Their bare branches would still support the thin maples. The maples entwined in their scaffolding would then be illuminated by sunlight and would bulk out and spread their branches. Gradually maple trees would occupy the spaces previously taken up by pines.

In my forest there are also "true" vines. They include wild grapes and Virginia creeper. All of them modify some of their leaves into grappling hooks to hang on. Other species of vines, such as English ivy, grow aerial roots that are designed to cling to bark. When climbing becomes a way of life, special holding devices evolve, to promote climbing. Almost any element of plant anatomy can be turned into a holding device for climbing. The holding device may be the stem itself, as it winds snakelike around its host. The climbing strategy can, through evolution, become ever more refined and specialized. However, many of its components are present to some degree in the normal growth patterns of most trees.

Virginia creeper has one of each set of paired leaves modified into grappling hooks for climbing.

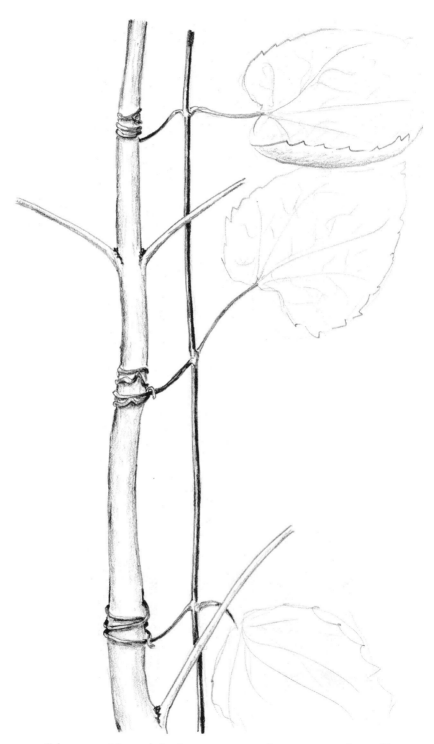

A wild grape with tendrils that wrap around a sugar maple sapling. Gaining support from the maple, the grape does not need to invest resources to make wood.

All trees reduce the investment for growing a thick trunk when they must grow rapidly upward to reach the light, as happens merely when they are in crowds. They can only afford to bulk out later to make a thick tall tree. In the meantime, the crowds provide lateral support, so less wood *needs* to be initially invested to make a thick trunk. If a limb accidentally rests on another tree experiencing less stress, then this limb can also stay thinner, and the tree can put even more effort into the top upward-reaching main shoot. In the stringent competitive economy of a forest, perhaps all tall straight trees result from an initial vinelike tendency.

TREE GEOMETRY AND APICAL DOMINANCE

Except during the nine months before he draws his first breath,
no man manages his affairs as well as a tree does.
—George Bernard Shaw

The glacially scarred ledges on the top of my hill are clad in a deep-shaded forest. Dark acid soil is covered with fallen brown needles and patches of luminescent green moss. Light brown cone bracts where a red squirrel has recently dismembered spruce cones are scattered in piles. Red spruces, the haunts of golden-crowned kinglets, abound here, and climbing up through a tangle of dead brittle branches in one of the red spruces I soon reach light. All around me I then see the pointed green tops of the individual trees. As I scan down the slope along the edge of the red spruce stand there appear the rounded forms of maples, poplars, and beech. Still farther down in the valley, almost to the stream, I again see pointy spirelike geometrical shapes, those of balsam fir trees. Scanning east across the valley, up onto the hogback of Holman Ridge, I see the great white pine trees sticking up like rows of ragged black teeth. These pines show up best against the orange sunrise, while the pointed spruces and firs become even more acutely pointed after a foot or so of snow has fallen and depressed the trees' branches. The more varied and soft round shapes of the different deciduous trees are outlined in early spring following flushes of the trees' flowers or emergence of new leaves in different hues of green.

Their individual shapes blend in summer into an otherwise uniform sea of green that is the top of the hardwood forest, and then in late fall they reemerge in flashes of brilliant reds, yellows, and purples.

When looking over the forest and seeing the beautiful and varied color-outlined shapes of the crowns of the trees I sometimes wander in my mind's eye to the bush veldt in Tanzania, where I once spent a year. There I saw acacias like flat-topped green umbrellas above the red sandy soil, with giraffes straining to reach their leaves. Then my mind travels back to Maine, where the beech leaves are arranged in layer after layer from near the ground to high up into the crown of each tree. Even though each individual tree is shaped by its immediate environment, every tree species still grows a distinctive form. Why and how is the tree's growth regulated to achieve its shape?

In my forest a major evolutionary constraint on tree geometry is snow loading and ice. On the African steppe a major constraint is the browsers like the giraffe. The young flat-topped acacias first shoot straight up, then they spread laterally and pack their leaves into a layer just out of reach of the ever-hungry giraffe. White pine trees have evolved to grow thick sturdy trunks that lift their shade-intolerant leaves less to deter browsers than to reach the sunlight far above the other trees. Pine trees can still have multiple layers of leaves because once a tree has reached above its neighbors, the lower layers can still be illuminated from the sides. The beech trees that grow on the north side of my hill among ash and white birch are much more shade-tolerant than pines. They can afford to have limbs and leaves even under other leaves. They grow slowly under diffuse indirect sunlight of the semishade occurring under their own leaves. They can therefore have multiple layers of leaves on the same scaffolding of trunk and limbs.

In the tropical rain forest there is no need to shed broad leaves seasonally. There, leaves can have a life that spans years, and the energy debt of leaf production does not need to be paid off every spring, as it must in an ash and other deciduous trees of northern forests. As a consequence, I suspect that many tropical trees can afford to be even more shade-tolerant than beeches and maples. Given the shade tolerance of many tropical trees, more layers of leaves can be packed in vertically, making the tropical lowland forest floor seem dark and gloomy in comparison to northern forests. The many "layers" of leaves in a forest can be occupied by the same or different species, but the shade-intolerant species will be the ones that, like white pines, must tower above the rest. As is obvious to anyone who keeps houseplants, however, the leaves of many species can also physiologically adapt to a range of light intensities (within limits), giving the tree flexibility.

Shade tolerance affects the shapes that are permissible. For example, a very shade-tolerant tree may spread out laterally in the understory to capture as much sunlight as possible. On the other hand, a "sun-loving" tree like a pine tries to shoot straight up, spreading laterally only after reaching above the other trees. The general outline of a tree's ultimate *potential* geometry can be seen as a mold that has been shaped by evolution. However, the immediate local environment determines the extent to which that mold is realized.

The role of physical stress in creating a balanced structure is not immediately obvious. For example, the thickest part of a tree, the lower trunk, needs to withstand the most physical stress. It is, of course, the oldest part of the tree, and the older it is the more it must, and can, support. The limbs reach out laterally and taper to ever-smaller diameters at the tips. The trunk and the limbs therefore both appear to be just the right thickness to support the appropriate amount of weight. The "perfect" tapering of trees and their limbs is in part the result of age, since the tip of a pine twig only has one growth ring of wood, whereas the bottom of the trunk that must support the whole tree has as many as the tree's age in years.

In addition to age, the amount of wood that is produced for support is also the result of stress experienced; stress affects the width of growth rings in trees as it affects the development of muscles in us. For example, if a young tree is bent over, then the next growth rings are thicker on what is now the underside of the trunk than on the top. The buildup of wood on the side where the trunk (or a limb) is compressed seems to act as a splint that gradually expands and pushes back in the opposite direction. As a consequence the tree will not be dragged down even lower by its own weight. Instead it will gradually right itself. Examining the width of the growth rings in the cross sections of stumps of trees, I noticed that they are sometimes as much as three times thicker on one side of the center than the other. Such asymmetry always means that the tree was leaning and then tried to right itself. In some giant tropical trees, such lateral buttressing by extra wood deposits may result in the growth of great

Stress affects wood buildup. In this double-trunk white pine tree, the now-drooping branch had been supported and it grew small and concentric growth rings. The other limbs of the same age grew thicker growth rings, especially near the lower portion of the branch.

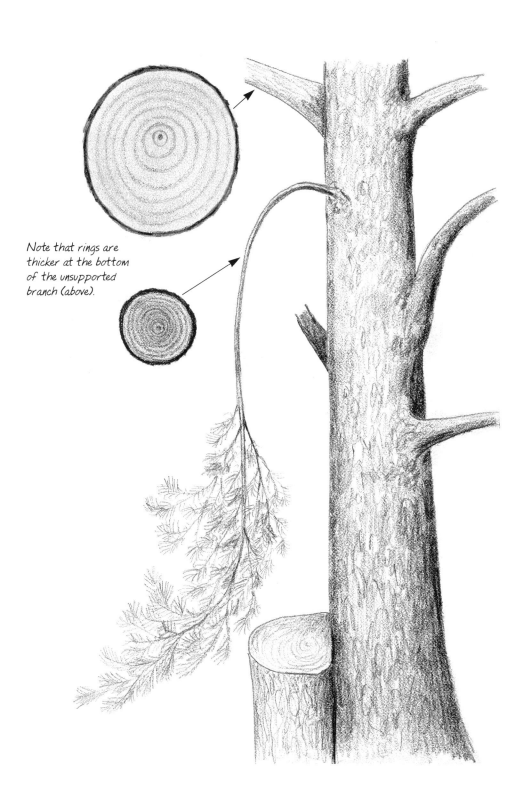

Note that rings are thicker at the bottom of the unsupported branch (above).

flanges reminiscent of the buttresses on medieval cathedrals. Trees evolved what the medieval cathedral builders invented hundreds of millions of years later. In tree limbs the greatest stress occurs in the "hinge" area, at the attachment to the tree. It is here that the growth rings of limbs are thicker on the bottom than on the top. A mechanism of selective wood deposition could help to brace the limbs up laterally.

In one of my pine trees, a "double-trunk" tree, a long lower branch had lodged itself on top of the neighboring trunk. Because this branch was supported by a "crutch," it had remained thin. Its growth rings were narrow and symmetrical about the center, unlike those of most branches, which require bolstering from the underside to keep them from drooping.

In plants, as in animals, growth is controlled by hormones. The translation of mechanical stress to a message that triggers the tree to grow more wood almost surely is regulated by hormones as well. Hormones activate genes. Genes regulate growth. Gross malformations result if certain genes are altered, especially when those that code for the hormones themselves mutate so that the hormones' actions are disrupted. Even a single mutation can radically change a tree's shape.

A mutation can sometimes be traced to a single kind of molecule. A mutation that affects thyroid hormone in animals yields stunted misshapen form, unless corrected with thyroid hormone supplements. Similarly, there are mutant dwarf bean plants that if treated with just one hormone, gibberellic acid, become completely normal. The implication is that the mutation in the bean plant results in the loss of the plant's ability to synthesize this specific plant hormone that is well known to promote cell elongation. Other hormones called cytokinins promote cell division. Together cytokinins and gibberellic acid then regulate size and form in trees, as well as in other plants. "Monster" trees, a mutant form that has been found in beeches, elms, and oaks, result when the tree fails to conform to its normal branching pattern. The result is gnarled, grotesquely short, bushy, and twisted trees.

Potentially, all that is needed to produce a prostrate, or creeping, "tree" is for the top shoot to fail to grow upward, a feature controlled by a single gene that codes for one hormone. Creeping forms can be adaptive in some situations. Arctic willows and dwarf birches that are evolved for living on mountaintops, for example, are only a few inches tall. Mere stunting by deliberate starving, bending, and pruning, as in bonsai, on the other hand, reflects a tree's flexible response to specific local conditions rather than a genetic alteration.

No trees show more *symmetrical* growth than firs and spruces. They are perhaps ideal subjects for finding out how local conditions affect ultimate shape. Those whose seeds chance to sprout on mountaintops tend to become gnarled and stunted like bonsai. However, the balsam firs in my forest are all tall and symmetrical trees. To see how growth determines their shape, I started to look first at the buds at the very top of the tree's leading shoot, because *all* lengthwise growth of trees occurs only at the tips, at just one, the center one, of this cluster of buds.

The top shoot of a fir, pine, or spruce in winter is the previous year's growth. This vertical shoot is topped by three to six buds arranged in a circle around one center bud, and each of these buds becomes a shoot the following spring. The one center bud becomes the topmost shoot for the following year. Ultimately it (normally) becomes the tree's trunk. The two to five buds that are arranged around the center bud become a whorl of branches that grow out laterally from the tree at nearly right angles.

Left: The lateral view of the top shoot (last year's growth) of a balsam fir tree, with newly opening buds at the top and sides.

Looking down on the "head" of a balsam fir tree's top shoot, showing the center bud that will become the trunk, and five lateral buds that will become limbs.

Looking down on the "head" of a balsam fir with buds (including six normally destined to become branches) just opening.

The branch tips have the same pattern of buds as the leading tip that becomes the trunk. One bud becomes the leader branch. As does the top shoot, only the central bud becomes leader of the branch. It is in this hierarchical control, with the central bud that becomes the tree's "leader" and the other buds that become lateral branches, that the tree's functional geometry resides.

The mystery is this: If by some accident the center bud of the tree's top shoot and all but one of the lateral (limb) buds, or even the central stem, are destroyed, then the one remaining limb bud or limb will alter its course to become the tree's new vertical "leader." How is the fate of a bud determined? The tree apparently "knows" the importance of having a leader and it will alter a lateral bud to become the leading stem, if needed. The tree practices priorities. The first priority is growth at the apex. Branches are secondary. If it did not set priorities, then it would stand no chance of reaching the light. If all the shoots were allowed to become leaders, then the tree would have more than one head or even become a spreading low bush that would quickly be overgrown by competitors. It would also be highly susceptible to collapse from even modest loads of snow and ice (provided it was not a creeping form like a lycopod).

Having two heads is much worse (for a conifer) than having just one. Almost invariably when two leaders vie for dominance and both grow on the same tree without one becoming subservient to the other, the tree suffers. In my pine grove there are a number of two-headed or two-trunked pines aged forty to fifty years. After every ice storm one or another of them is eliminated as one side of a tree crashes down, leaving the other side with a large gash on its side. The remaining tree is also destined to crash. The early mistake that resulted in one twig not yielding to the other is not easily corrected once the tree matures and is subjected to mechanical stresses.

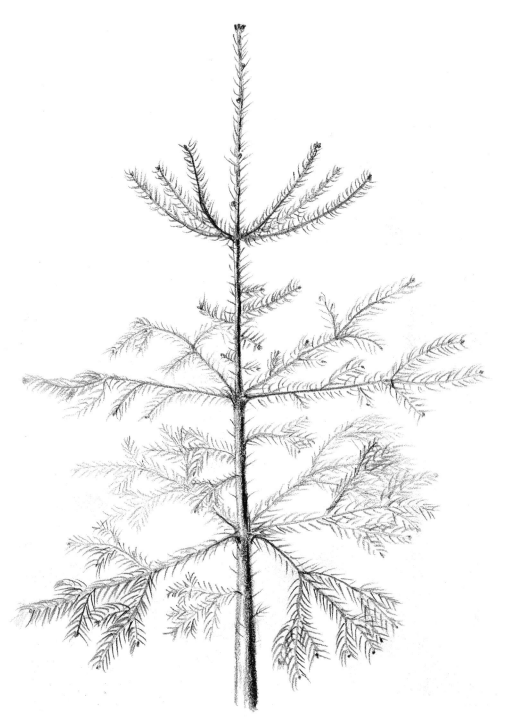

Balsam fir tree, showing each of the last four years of growth, as can be
seen by the branch patterns.

However, as we shall see, the tree has intricate biochemical mechanisms to ensure that there is (usually) just one leading shoot.

In most cases apical dominance works, and we tend to take it for granted because of its success. However, the mechanism that normally maintains apical dominance sometimes breaks down as a result of accidents. In my pine trees, there are two main causes for breakdown of apical dominance. The first is a small beetle, from the weevil family. This beetle lays its eggs into the freshly growing terminal shoot of white pines, provided they are growing in sunshine. The beetle larvae feed inside the tissue and kill it, effectively decapitating the tree. The tree responds to decapitation by trying to regrow a new head from its various lateral twigs. Unfortunately, once stimulated to grow a new head, it often ends up growing several.

Before the weevil larvae attacked, the tree had already made a "decision" about which shoot would lead. Now, after the attack, it must choose again. The problem is that all of the lateral twigs below the dead leading shoot are equally capable of making the new head that becomes the tree's trunk, and all appear to simultaneously get the message that a new leader is required. All respond to fill the void. Being in sunlight they all grow quickly, and soon several shoots are long. The tips of the shoots are far apart and no longer appear to communicate one with another. It is as if they do not yet know that a replacement tip is active. Hence none is suppressed. The more two heads have grown independent of each other, the more they become independent, and a "bush" tree, growing in all directions, results. Such trees, called cabbage pines, stand no chance against competitors that grow straight upward, nor do they provide useful lumber. Several of these trees at the edge of my clearing lose great limbs after almost every ice or snow storm.

Apical dominance, one of many phenomena responsible for the

shape of trees, was first discovered in other plants. It was originally discovered by clever experiments done by Charles Darwin and his son Francis who reported their experiments in a book published in 1881, *The Power of Movement in Plants*. The Darwins experimented with grass seedlings, which, like tree seedlings, normally grow straight up rather than horizontally. When illuminated laterally, the grass seedlings will grow horizontally—they bend and grow toward the light. However, when the tops of laterally illuminated seedlings were enclosed in a small black cap, they continued to grow straight up even when illuminated from the side. The Darwins concluded that something in the growing tip "sees" the light and it then causes the seedling to bend toward that light source.

Nearly half a century later, plant physiologist Fritz W. Went began to unlock the biochemical mystery of how the plant responds by growing toward light, as well as maintaining apical dominance. He isolated the growth hormone indoleacetic acid (IAA) from the growing tips of shoots and determined that this hormone diffuses downward from the shoot tip inhibiting the buds below it even while simultaneously stimulating elongation in the cells of the shoot itself. Significantly for the bending response, the hormone also moves from the illuminated to the darkened side of the shoot. Thus, the cells at the shoot's darkened side elongate and the shoot bends toward the light.

At least superficially, it seemed as though the same mechanism of bud inhibition caused by a substance in the growing tip applied to oaks, beeches, ashes, and maples. On May 23, I found fresh leading shoots of these trees already grown to more than six inches long, while most of the lateral buds had not yet even opened! However, the same pattern didn't seem to hold true for conifers. On the same date I found that most of the *center* buds in the top shoot of fir trees were just barely opening, whereas the *lateral* three to six buds

White ash top or terminal leaf bud opening while the two small lateral buds just below it (see arrow) are still dormant.

already had vigorously elongating shoots. Was there apical suppression rather than apical dominance? To me, this didn't make sense, and I was pleasantly surprised. I no longer had even a hypothetical explanation as to how the then-smaller central bud might *later* become the tree's trunk.

A month later I again examined the tops of the small firs behind my cabin. At that time the annual new growth of wood was nearly completed and the central fir shoots had then grown straight up one

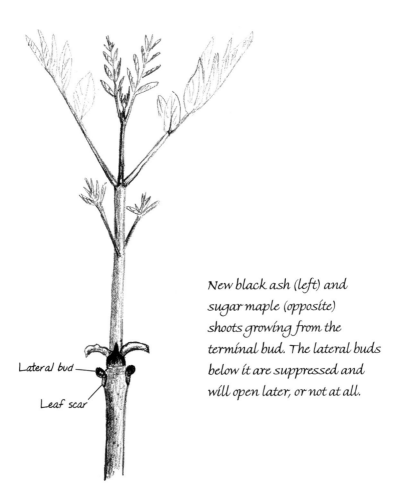

Lateral bud

Leaf scar

*New black ash (left) and
sugar maple (opposite)
shoots growing from the
terminal bud. The lateral buds
below it are suppressed and
will open later, or not at all.*

or two inches in shaded trees and up to sixteen inches in trees in sun-shine. In both cases the central shoots had now overtaken the lateral ones. The lateral shoots (from the *lateral* buds) had now for the most part also bent to the sides at the normal angle that limbs come off the tree. What had prevented them from bending *up*?

I snipped off the center shoots on some of these trees. On others I removed all of the shoots except one lateral shoot. I wanted to see if the central shoot had any role in the growth of the others. I did not expect to see results for a year or two, because the current year's

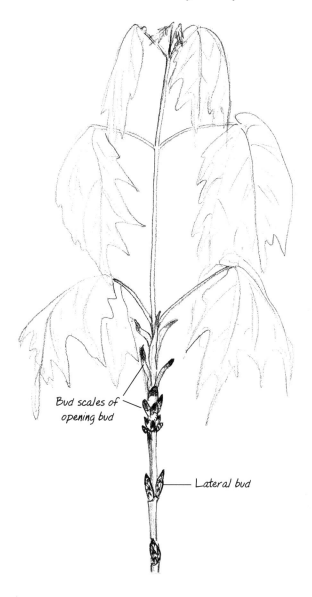

Bud scales of
opening bud

Lateral bud

growth was nearly done. So I also looked for naturally occurring amputations. Deer, for example, occasionally snip off the tops of young fir trees in the winter, so I examined the results of their activity.

I found several young fir trees whose leading shoots had been snipped off the year before. In all cases one of the lateral twigs had then bent upward. This upward-directed twig became more elongated than the other lateral shoots from then on. The trees do indeed somehow have a physiological capacity to pick a leading shoot from among competing shoots. But how do they do it? My observations only point out mysteries. Systematic research would reveal even more. I find that reassuring.

TIME TO A TREE

There is something even in the lapse of time by which time recovers itself.

—Henry David Thoreau

The morning of October 28, 1996, was cool and overcast. A wind blew from the northeast and I expected the first snow of the year at any time. I was restless and left the cabin for a walk in the woods.

The trees had just shed their leaves, revealing many abandoned bird nests. A goldfinch had left its plantdown-tufted nest in the top branches of a sugar maple growing alone in the clearing. Just behind the cabin I now saw a robin's mud-lined nest and two nests of cedar waxwings decorated with lichen. Then I saw two red-eyed vireo nests artfully suspended with strips of white birch bark in the forks of horizontal maple twigs. The nests must have been completely covered with broad green leaves in the summer, because I had searched in vain here earlier where the birds sang, scolded, and lived. All of these birds had timed their arrival in the spring to that of insects, and they had waited to build their nests under the cover of new leaves.

The winterberry bushes at the edge of the woods across the clearing from the cabin were visible from afar. I saw them now as a brilliant red patch because their thousands of plump berries contrasted against gray twigs. The fruit could last on the bushes for four more months. These berry bushes had sprung up unexpectedly a few years ago, growing in the only wet open spot around. While removing brush from the clearing I had at first wanted to cut them out. But I was soon glad I had left them.

A flock of robins arrived soon after I returned to the cabin. I saw about a dozen of them perched in the trees. More of them were running in jerky, short sprints on the matted grass in the field. It was hardly a great crowd, but it was part of the great throng of the fall migration. For the moment I did not pay them much attention, because I was still engrossed in examining the crop contents of a grouse. My friend Glenn Booma had left them (in a plastic bag), along with a note, on my table during his last visit to the cabin. The note said "Bernd—check this out!!" I did. The crop contents from this one grouse contained 52 striped maple seeds, 2 oxalis leaves, 7 quaking aspen leaves (presumably from young shoots that sometimes grow from roots late into the season), 54 quaking aspen buds, and 11 red berries. Might these berries be winterberries? To find out, I walked back down to the end of the field to the winterberry bushes to compare. To my great surprise, all of the bushes were now totally bare. They had been denuded during the past hour. Not a single berry was left. Within minutes the robins were gone, too. They would now likely scatter winterberry seeds behind them in a sinuous path that might stretch over a hundred miles. The winterberry had picked a good time to present its fruit, as had all of the berry trees. The shadbush trees at the other end of the clearing were pollinated in early May by bumblebee queens just out of hibernation, and they had a big crop of purple berries ripened in June, ready to be eaten by cedar waxwings who had just returned from migration. In June the

Sequence in the unfolding of a beech bud. In about four days the bud unfurls, the twig extends, and the pleated leaves spread and grow, after a dormancy of ten months.

waxwings were flying on schedule from one clearing to the next searching for suitable nest sites and meanwhile dispersing shadbush seeds in their wake. Trees apparently also have schedules. They have times of flowering suited to make use of insects and wind for pollination, and their fruiting is scheduled so as to enhance seed dispersal by birds and wind.

Just as trees time their flowering and fruiting to maximize pollination and seed dispersal, periods of dormancy are also part of a tree's cycle. The trees in my forest already have buds in July, and these buds formed at the ends of twigs that grew from buds opened in May. These *new* buds almost invariably stay dormant through the summer, fall, and most of the winter. Buds on twigs will not open in the warmth of a house in December. However, there comes a time two to three months later when they will respond to warm temperatures and open. They will open because the time "scheduled" for dormancy, and somehow measured by the twigs of the tree, has passed.

White ash in late July with new buds.

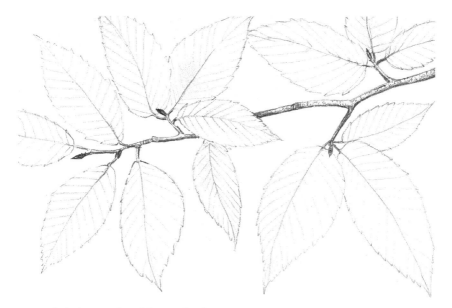

Beech in late July with new buds.

The dormancy of the buds through winter is adaptive, because otherwise a stray warm day might induce leafing out prematurely. However, dormancy through summer and even into autumn (as in ash, maple, beech, and oak) seems strange, because there is still much time left for growing. Nevertheless, occasionally a single bud of ash, maple, or oak breaks rank, opening in July, almost a year ahead of schedule.

I first noticed these "outlaw" buds in the two young vigorously growing oaks in my clearing. Their new twigs had grown to full length, and the leaves were mature and dark green with new buds already in place. Suddenly in July the ends of a couple of the new twigs had the baby pink leaves that red oaks have when they are just out of the bud! *One* of the cluster of terminal buds had broken dormancy. Curiously, only on two new twigs had one of the buds broken dormancy. The many thousands of others remained inactive. They were biding their time until next spring.

I checked the twigs again closely in early October when the leaves were still dark green but starting to turn purple brown. There was now no longer a difference between the color of the May and the July leaves. Both had adhered to the same schedule of color change. The two new shoots had grown only two to three inches, but they each held a full complement of five and eight leaves. And the new terminal shoots had new buds.

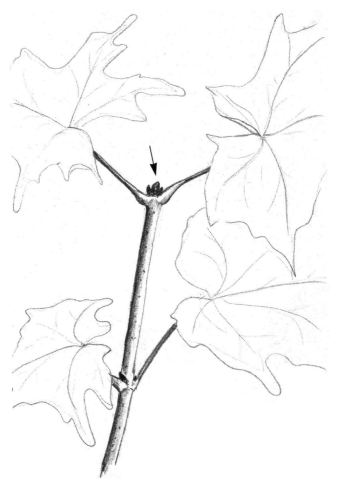

One sugar maple bud that broke dormancy in late July.

Why had several buds not marched to the same drummer as all of the other buds? Why had they taken the jump almost one year ahead of them? It was as if they had tried to skip one whole year of maturation. Would they get away with it? I suspected there would eventually be a cost, otherwise all buds would skip dormancy. Might the cost be twig death? Do the buds need the time to get prepared to survive freezing? I put small wire tags on the two twigs and

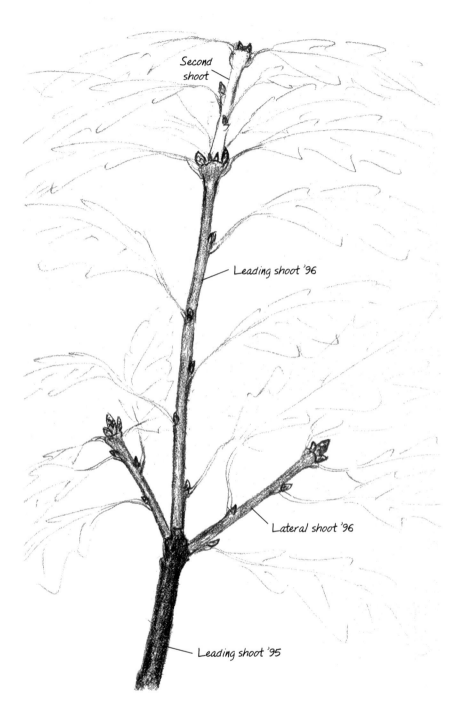

Second
shoot

Leading shoot '96

Lateral shoot '96

Leading shoot '95

A red oak twig whose terminal bud broke dormancy in late summer to grow a second short shoot from a new bud.

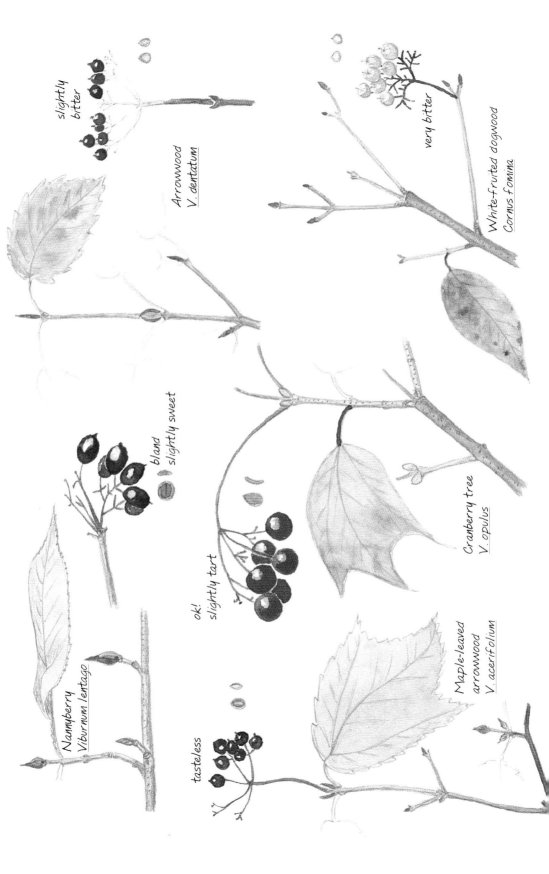

slightly
bitter

Arrowwood
V. dentatum

very bitter

White-fruited dogwood
Cornus fomina

bland
slightly sweet

ok!
slightly tart

Cranberry tree
V. opulus

Nannyberry
Viburnum lentago

tasteless

Maple-leaved
arrowwood
V. acerifolium

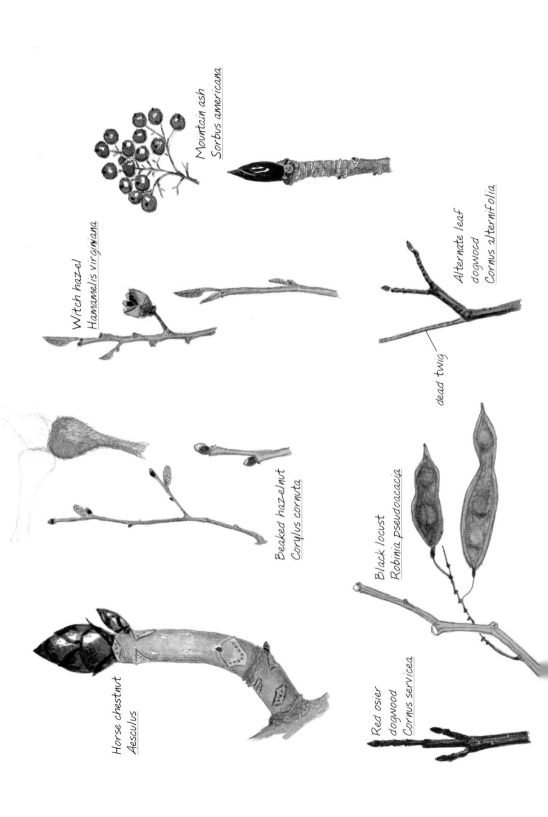

Mountain ash
Sorbus americana

Alternate leaf
dogwood
Cornus alternifolia

dead twig

Witch hazel
Hamamelis virginiana

Beaked hazelnut
Corylus cornuta

Black locust
Robinia pseudoacacia

Horse chestnut
Aesculus

Red osier
dogwood
Cornus servicea

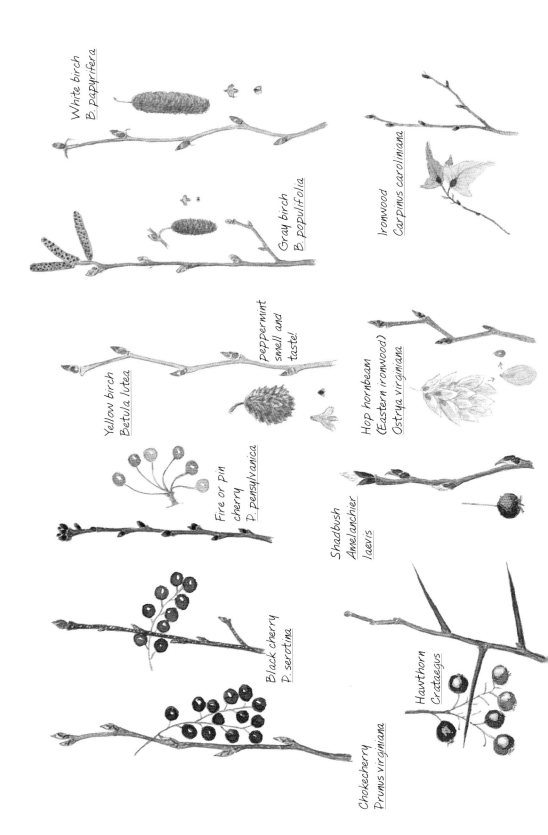

White birch
B. papyrifera

Gray birch
B. populifolia

Ironwood
Carpinus caroliniana

Yellow birch
Betula lutea

peppermint
smell and
taste!

Hop hornbeam
(Eastern ironwood)
Ostrya virginiana

Fire or pin
cherry
P. pensylvanica

Shadbush
Amelanchier
laevis

Black cherry
P. serotina

Chokecherry
Prunus virginiana

Hawthorn
Crataegus

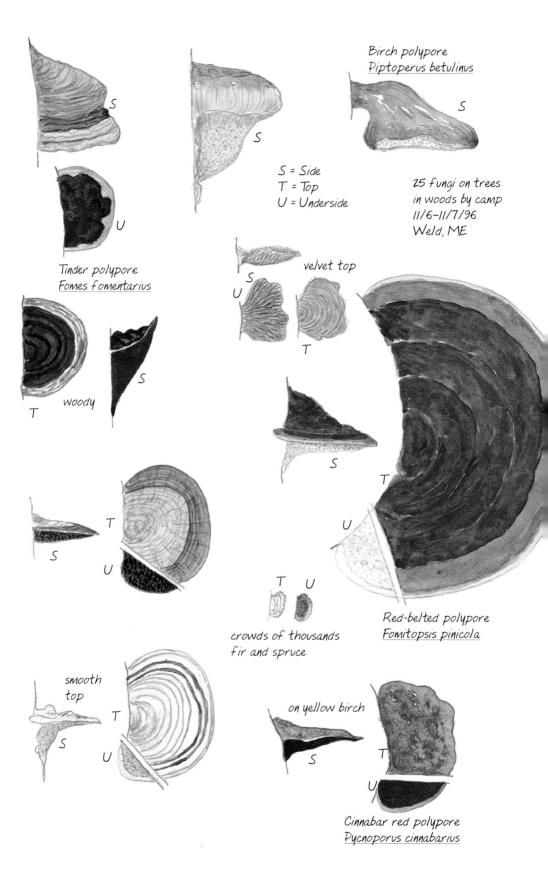

Birch polypore
Piptoperus betulinus

S

S = Side
T = Top
U = Underside

25 fungi on trees
in woods by camp
11/6–11/7/96
Weld, ME

S

U

Tinder polypore
Fomes fomentarius

velvet top

S
U

T

T

woody

S

S

S

T

U

T   U

crowds of thousands
fir and spruce

Red-belted polypore
Fomitopsis pinicola

smooth
top

T

S

U

on yellow birch

S

T

U

Cinnabar red polypore
Pycnoporus cinnabarius

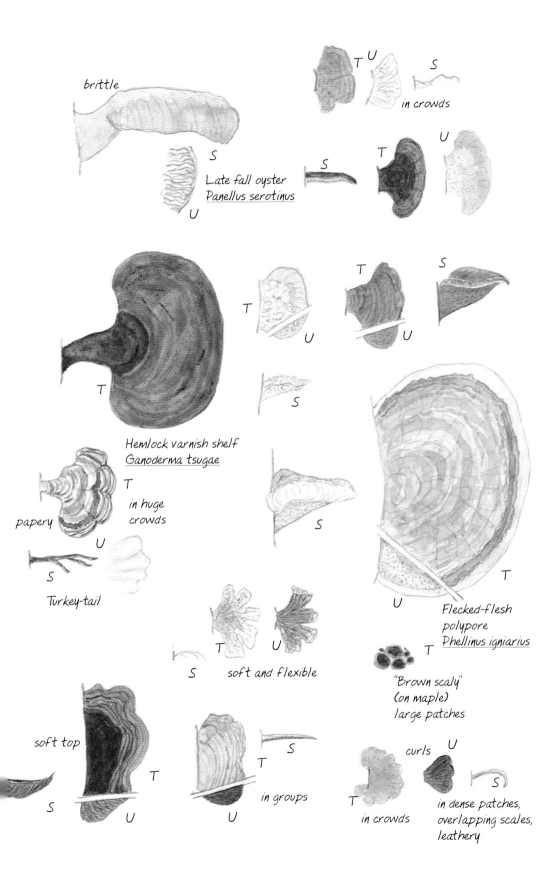

brittle

T U

S

in crowds

S

Late fall oyster
Panellus serotinus

U

S

T

U

T

U

S

T

U

S

Hemlock varnish shelf
Ganoderma tsugae

T

in huge
crowds

papery

U

S

Turkey-tail

S

T

U

soft and flexible

U

T

Flecked-flesh
polypore
Phellinus igniarius

T

"Brown scaly"
(on maple)
large patches

soft top

S

T

U

S

T

U

in groups

curls

U

T

S

in crowds

in dense patches,
overlapping scales,
leathery

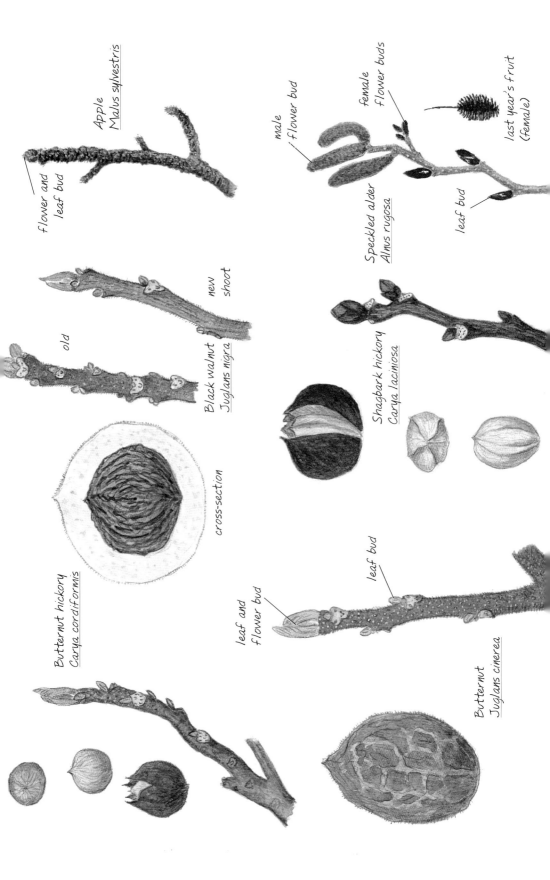

Apple
Malus sylvestris

flower and
leaf bud

male
; flower bud

female
flower buds

leaf bud

last year's fruit
(female)

Speckled alder
Alnus rugosa

new
shoot

old

Black walnut
Juglans nigra

Shagbark hickory
Carya laciniosa

cross-section

Butternut hickory
Carya cordiformis

leaf bud

leaf and
flower bud

Butternut
Juglans cinerea

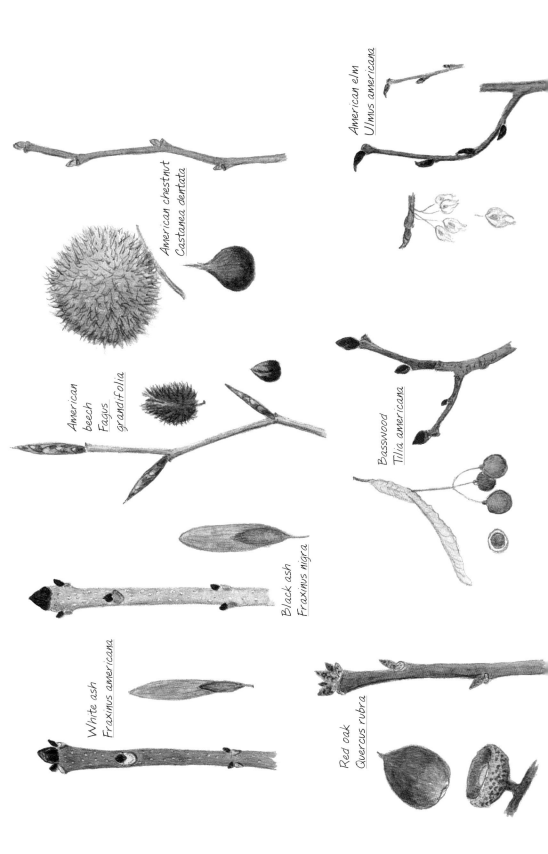

American elm
Ulmus americana

American chestnut
Castanea dentata

American
beech
*Fagus
grandifolia*

Basswood
Tilia americana

Black ash
Fraxinus nigra

White ash
Fraxinus americana

Red oak
Quercus rubra

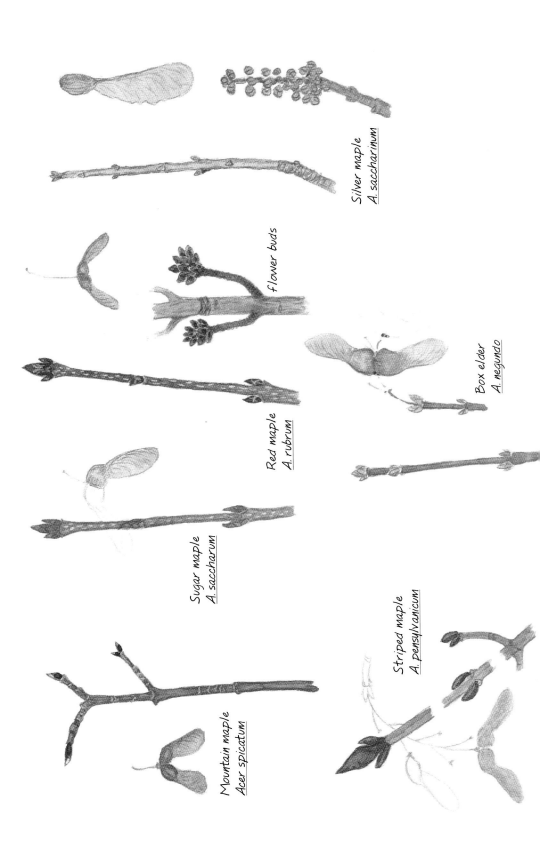

Silver maple
A. saccharinum

Flower buds

Box elder
A. negundo

Red maple
A. rubrum

Sugar maple
A. saccharum

Striped maple
A. pensylvanicum

Mountain maple
Acer spicatum

checked them on June 1, 1997. To my great surprise the twigs looked healthy, and the buds had opened to produce healthy new shoots with new opening leaves.

Time schedules are most strict for flowering and fruiting. The alders flower in early April and the basswoods in mid July. The shad-bush ripens its berries in early June and the winterberry fruits in late fall. I can make observations and then speculate about these schedules on the assumption that they came into being, and persisted, because they uniquely served the tree of each species to reproduce itself better in the particular environment that it grows in. Beyond that, I wonder *how* the trees can adhere to the proper time for each of their many schedules. Do they have a clock from which to read time?

Biologists have intensively studied how organisms measure time for more than half a century. They have learned that virtually all types of organisms, from protozoa to bean plants, bees, and people, have "biological clocks" operating within each cell or specific group of cells. These "clocks" are primarily "circadian" clocks; that is, they measure time in approximately twenty-four-hour cycles, as if off a watch that is set to run in repetitive approximately (i.e., *circa*) twenty-four-hour cycles. They *run* in the absence of external cues as any mechanical watch does, but external cues called *zeitgebers* (time givers) *set* the "clocks." The clocks can be *re*set to start their twenty-four-hour cycles at any time, by environmental signals such as the moment of light on, or light off. These environmental zeitgebers routinely reset the biological clocks in the same way that we period-ically need to reset cheap wristwatches that run, say, ten minutes late (or fast) each day. However, we cannot speed up a biological clock—or a wristwatch—or slow it down. We can only reset the dial. Do trees have circadian clocks and if so what activities are determined by them?

I do not know the answers. At least superficially, there seems to be little need for trees to keep track of circadian time. Twenty-four-hour cycles may be necessary for some tropical trees to synchronize nectar production and flower opening with the daily or nightly schedule of a specific pollinator. However, I doubt that such cycles exist in most trees of the northern forest. The schedules that northern trees do adhere to are those of the *seasons*.

Most organisms living in a regular seasonal environment have adapted to anticipate the seasons. Woodchucks lay up fat long before it is time to hibernate, just as I chop my firewood long before I first need it in November, because it needs to dry through the summer. Similarly, northern trees where the growing season is short must have their buds nearly fully prepared by fall, so that they can quickly unfurl them at the first warmth of spring.

Changing temperature could be one cue trees use to predict the change of seasons. However, the temperature often changes wildly and is therefore an unreliable cue. The amount of sunshine changes seasonally, but the *amount* of sunshine is also an unreliable cue because there can be cloudy weather for weeks on end. Nature's most reliable cue of the present season (and hence the coming one) is photoperiod—the relative number of hours of light and dark in a twenty-four-hour cycle.

Mysterious as it may at first seem, many plants and animals respond to seasonally changing photoperiod, using their *circadian* clocks to measure. The variable they measure is the time interval between dawn and dusk, or that between dusk and dawn. This requires the circadian clock. I suspect, therefore, that although most of the trees in my forest do not show externally visible twenty-four-hour *schedules*, they could nevertheless still consult a circadian, or twenty-four-hour, clock to keep track of the seasons.

Time to a tree is also marked by more irregular intervals between events than the circadian and annual cycles that we share with them. A tree starts out as we do, as the union of two haploid gametes, the egg and the sperm. Cells divide again and again to become embryos. An infinitesimally small number of embryos become seedlings, and out of these, a tiny subset become trees. Time may stand still for decades in some seeds that enter dormancy until conditions for growth become suitable. Time may stop for decades as the seedling exists in the shade, garnering just enough energy for survival but not enough for growth. Time speeds up as sunlight is reached and the tree explodes in sudden growth and then proceeds along its trajectory to reproduction, senescence, and death. In some trees, like the gray birch and balsam fir, maximum life span is usually shorter than our own. In many others it is close to ours, while in a few trees, including the bristlecone pines, a life span of four thousand years is not impossible.

Bristlecone pines grow extremely slowly because they live in a cold climate (the White Mountains of California) and because they have little water and few nutrients. They nevertheless stay a step or two ahead of decay and death because the climate also dries their deadwood. It takes them thousands of years to experience the growth, and life, that a white pine in Maine experiences in two hundred summers. Trees must be growing to be alive, but different species grow, and therefore live, at very different rates. Thus, even to a tree, both time and life are relative.

SEX IN TREES

There is more of good nature than of good sense at the bottom of most marriages.

—Henry David Thoreau

One day in mid April, the alders down by the brook unfurled their hard, tight, purple catkins. Overnight they were transformed into long, loose dangling tassels that trembled in the least bit of a breeze. Wisps of yellow pollen flew off like puffs of smoke. The brook, aptly named Alder Stream, roared full of tannin-stained yellow water collected in rivulets off the hill still under thick banks of snow. Slightly above the alder thickets under the bare poplars, the beaked hazelnuts were also unfurling their male catkins. As in the alders, the female hazelnut flowers are on the same twigs. But I see these tiny delicate purple wisps only when I look closely. Windblown pollen grains find them.

The alders, hazelnuts, poplars, willows, sugar and red maples, beeches, ashes, and oaks each shed their pollen in a few days. All are done blooming when the first leaf buds open from the beginning of May and on into June. The insect-pollinated trees, the basswoods and chestnuts, wait to flower until summer, weeks after they have

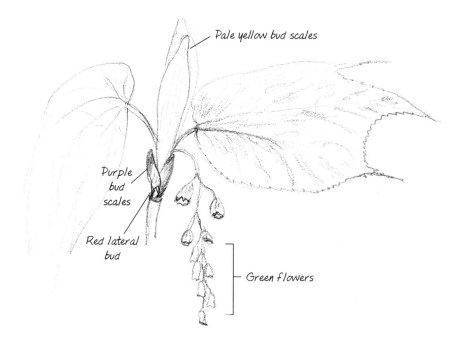

Pale yellow bud scales

Purple bud scales

Red lateral bud

Green flowers

The terminal bud of a striped maple opens to reveal both leaves and flowers. The flowers are small and green, and they are both wind- and insect-pollinated.

leaves. Each species has a schedule, and that schedule is especially strict in matters pertaining to sex.

Flowers are sexual organs, and flowering is not sex but the preliminaries. A great deal of investment goes into sex, and the mechanisms that animals and plants have evolved to achieve sex are a source of wonder and beauty. Without sex there would be no flowers and no bees, nor would there be brightly colored birds and birdsong. There would be no alluring scents that female moths give off at night that attract males from a mile or more downwind. There would be no flashing of fireflies through the trees on hot July nights, nor would there be evening serenades of crickets and katydids in August.

Scents, sounds, and colors often combined in fantastic forms have the purpose of attracting mates. But trees can't be attracted. They are rooted in place.

In order to achieve sex, each male tree (or the portion of it that is male) sends forth millions of tiny pollen grains. These structures contain sperm, but they are not sperm. Instead, they are analogous to whole male reproductive tracts. When one pollen grain lodges on the stigma of a flower—the receptive part of a compatible female reproductive tract of another tree—it sends forth a penislike tube that grows rapidly, thrusting into the plant's ovary. Only then do the sperm descend through this self-made tube into the ovary to fertilize the eggs. However, the female is not passive. She chemically recognizes the proper pollen and permits the growth of pollen tubes only for mates she chooses. The plant's own pollen is usually rejected or accepted only when none other reaches her.

There are several ways whereby a tree's pollen grains may find appropriate ovaries. The forte of the angiosperm plants is their ability to harness intermediaries to carry the pollen. Apple trees, for example, have harnessed bees by tailoring the flowers' shapes, colors, scents, and food rewards to suit bees. Apple flowers are perfumed and colored bright pink. They stand out like billboards to bees and advertise to them from afar, inviting them to come sip sweet nectar and take pollen. As bees feed on the flowers' nectar they become dusted with pollen. They scrape most of this pollen from their bodies and collect it to feed their larvae back in their hive. However, some of the pollen remains on their body hairs, and it inadvertently gets deposited on other apple flowers on other trees that the bee also visits. The bee thus becomes a long-range mating agent for trees rooted in place. In some other plants, such as saguaro cactus trees, the much larger white flowers open only at night and provide much more nectar than apple flowers. They are shaped so that the fur of specific bats that are adapted to eat nectar and pollen gets dusted with pollen, and this pollen then gets deposited in flowers of the next plant the bat visits. Many other plants are tailored in shape, color, scent, and the food they offer to attract moths, butterflies, flies, beetles, or birds. In turn, the flower bats, hummingbirds, sphinx moths, and bees are highly dependent on the plants for their own livelihood and neither organism can now exist without the other. Through evolution, flowers made these animals what they are, and the animals in turn made flowers.

Almost all of the trees in my forest use wind, rather than bees or other animals, to transfer their pollen. Wind has drawbacks. It is unpredictable. It may be absent when the flowers bloom, or it could blow from the wrong direction. Insects, however, are there to be beckoned at any time, and they can come from any direction. In comparison to a bee flying directly from flower to flower, using the wind to send pollen or to receive it from a neighbor might seem like a big gamble for achieving sex. As in all gambles, however, there are ways of improving the odds. Trees improve their odds of fertilization by dispersing astronomical numbers of pollen grains. More than five million are released from a single birch flower or catkin, and each tree has hundreds of thousands of flowers. Female wind-pollinated flowers have enlarged sticky stigmatic surfaces that aid in catching pollen grains out of the air. In all of the deciduous wind-pollinated trees (ashes, elms, beeches, sugar and red maples, oaks, birches, willows, poplars, and aspens), the narrow window of time when sexual unions occur is very early in the spring before the leaf buds open. If flowers waited until after the leaves unfurled, then they would be partially shielded from the wind and there would be less chance for successful transfer of pollen between flowers.

Animals search out flowers even within foliage. After having been rewarded with food from one kind of flower, most animals search persistently for others of the same kind, identifying them by scent, color, and shape. Male and female flowers must be similar in appearance; otherwise the pollinator might fly to flowers of one sex and not the other. Having both sexes on the same flower, such as in the apple, is one common solution that helps solve this problem (so long as the plant also has means to prevent self-pollinating). The sexual system of an apple tree thus differs from that of an animal in that male and female roles are usually not mutually exclusive within one individual. Most plants have little or no sexual identity. An apple blossom is both male and female. It gives and receives pollen. Having both sexes on the same flower, apples have so-called perfect flowers, as opposed to "imperfect" ones (either male or female), as are found on alder, hazelnut, or maple, for example.

Wind-pollinated plants do not have showy flowers. In them the appearance of male and female flowers is irrelevant and the sexes are usually on separate flowers. Sometimes there are even whole trees that have either all male or all female flowers. In willows, for example, some individual trees have only male flowers and others have only female flowers. In American white ash trees and red maples there are (usually) separate male and female trees. Two of the three American chestnuts I have planted have so far produced both male and female flowers on the same tree. However, one produced primarily male flowers. Only the female flowers produce nuts. On hazelnut, alder, beech, and oak, separate male and female flowers are always on the same tree, and to achieve cross-pollination these trees must avoid fertilizing their own ovaries with their own sperm. Trees avoid self-fertilization and achieve cross-pollination in various complex ways. Given all the large investment, and all the complex and

wonderful mechanisms that have gone into achieving sexual reproduction, one wonders: What is the point of it?

The point of *reproduction* is to pass on as many of the individual's
genes as possible. If a tree's flowers were self-pollinated, then all of
its seeds would contain only its own genes. One that must be cross-
pollinated automatically halves the number of its genes that it packages into each seed, since the other half comes from an unrelated
other parent. Cross-pollination from the perspective of reproduction is thus not a solution. It's an immense problem. What does sex
do for the tree? Why bother with it? Why should a tree cross-polli-
nate when, by so doing, it reduces the number of genes it passes on
relative to what could be achieved by self-pollination?

Sex is a mechanism that introduces genetic variety into offspring.
Sex dissolves old genetic combinations, scrambles them, and randomly reassembles them into many new mosaics. It's a little bit like
having two similar cars (originating from two factories), taking them
apart piece by piece, putting all the pieces into one heap, duplicating
the pieces many times, and then picking appropriate pieces of each
kind randomly to reassemble them again into numerous new cars.
Each of these cars will have different percentages of parts in different
combinations from the two parent lines.

Every one of these new cars is apt to be slightly different. Some
will perform better and some will perform worse in each varied situation. However, genetic variety increases the odds that one or
another of these "offspring" will succeed in a changing or variable
environment relative to that of the "parents." Of course, the two
original cars can't be too different—say a VW "bug" and a Mack
truck—else the parts won't mesh and then none survives.

For trees and other plants as well as animals, diseases may be one
of the greatest variables of the environment. Right now, for exam-

ple, blister rust is devastating the white pines and Dutch elm disease the elms. Both of the fungi that cause these diseases are highly effective because they have fine-tuned their biochemistry to that of potential hosts. If they eventually specialize to attack a clone rather than a sexually reproducing tree, then *all* "individuals" of that clone would be equally vulnerable. Since each pine and elm tree is the product of a sexual union, however, there is more chance of individual immunity or resistance, offering them an evolutionary escape mechanism. Given sufficient isolation and time, potentially a whole new population could arise from a new variant. The rewards of sex in trees are not immediate. They are strictly long-term.

In the short term there are often situations when sex is disadvantageous to trees. Perhaps the greatest drawback is when potential mates are not within the trees' sexual reach. Since trees cannot travel, they face a problem. I could see it in my chestnut trees. In the fall of 1996 I examined my American chestnut trees and found that they had both male and female flowers. So far so good. The trees had even produced the typical round, spiny fruit of chestnut trees. However, inside the fruit were only the empty shells of typical-looking brown nuts. I suspect therefore that American chestnut trees don't self-pollinate. The flowers had probably not been cross-pollinated, so no embryos grew although receptacles for them did. Perhaps there were not sufficient numbers of suitable sexual partners nearby, or maybe the critical pollinating insect required by chestnuts was absent. Similarly, I found the ground strewn with sterile nuts under a lone black walnut tree (on a farm in Vermont). I eagerly picked up enough walnuts to fill a grocery bag. Unfortunately, each of thirty-five nuts I cracked was empty. The flowers from which they had grown probably had not been cross-pollinated for lack of a nearby partner.

I have often wondered why most of the trees in my forest use the wind to transfer their pollen and in many cases to disperse their seeds. In tropical forests, the trees hitch both their pollen and seed transfer almost exclusively onto animals, and the animals in turn come to depend on them. Especially in the tropical forests, trees can scarcely be thought of as separate organisms at all. They exist as parts of an unfathomable complex web where each species is linked to others in growth, reproduction, and sex.

Nobody has "the answer" to why northern trees are mostly wind-pollinated rather than animal-pollinated. There are probably many answers. Maybe there is not enough wind in tropical forests, or not enough animals in northern forests. Perhaps there are fewer animals in northern forests simply because the trees don't offer nectar rewards because they use the wind. Cause and effect often become blurred.

Tree distribution is undoubtedly important in determining the kind of pollination system that is needed to accomplish the job. A one-hectare plot (about two and a half acres) of Borneo rain forest selected at random contains about seven hundred species of trees, and the trees of any one species therefore tend to be *isolated* one from another. In contrast, in my forest, a one-hectare area selected at random would likely contain only one-hundredth the variety of trees, and many trees of a single species grow in dense crowds. Different sections of the southern slope down to Alder Stream are dominated respectively by quaking aspen, red maple, and sugar maple. The top of the ridge grows mainly red spruce, while on the western slope of the hill balsam fir and white birch predominate. The alders grow in patches along the side of the brook. In most cases, the individual trees of the same species touch or almost touch each other. I can hardly differentiate individuals. I doubt that pollinators do so. The trees are an *anonymous* crowd; they are not individually labelled by unique sign or location.

Looking at the pollinators' world in human terms, it is as if a street had many unlabelled, identical-looking, and identically scented restaurants all randomly intermingled. Some of them serve ample food of superior quality and some of them serve very poor food or none at all. To enter any one you have to pay at the door, and customers must visit several locales each evening to be satiated. As long as customers feel that on average they get suitable food they'll take chances and keep on coming to that group of anonymous restaurants. The restaurateurs, however, are individuals, and they are in intense competition with one another. In order to stay in business many of the more shrewd or dishonest slyly provide less food. Others must follow suit, with the end result that the customers eventually stop coming to eat on that particular street, because on average they keep paying the same and getting less. Substitute plants with nectar-rewarding flowers for restaurants, and bees for people, and we have the same situation. The plants get a payoff not from having the nectar eaten out of their flowers, but from having the insects enter their flowers, which pollinates them. When the customers no longer enter their flowers, plants need to devise a new mechanism of pollination. Could sex in the wind be the solution?

Apple Trees

Oh, give us pleasure in the orchard white,
Like nothing else by day, like ghosts at night
————Robert Frost

I lived for a while on a small farm near my forest. This farm was far out on a dirt road, and in spring that road became a series of mud wallows. Sugar maples grew alongside the road, and the fields were surrounded by forest. The tree I remember most was an apple tree next to the chicken shed. In summer, this apple tree shaded the rusted farm machinery and the whetstone that we turned by pumping a treadle with one foot. The whetstone had once been used to sharpen the scythes that were swung every summer to cut brush to keep the forest at bay, as it always threatened to creep back in from the edges along the stone walls.

One sunny May morning every spring the apple tree would erupt in a blaze of pink and fragrant bloom. Honeybees came in a steady stream to and from the box hive of pine boards and basswood honey frames that sat on the windowsill in the house attic. They swarmed around the apple tree, humming without pause. On such days when the bees came out of their hive, we didn't need to be threatened by the Hereford bull to make climbing the tree inviting. Sitting on a sturdy branch among the blossoms, we saw not only these honeybees but also burly fuzzy bumblebees that had emerged from their underground nests in rotten stumps. Occasionally we heard a low hum and we'd look to see a jewel of the bird world, the ruby-throated hummingbird, resplendent with its metallic reflecting ruby throat patch and sparkling green back.

Within days all of the pink petals fell to the ground like drifting snow, to vanish as quickly as they had arrived. Small green apples soon swelled on the ends of the twigs in the warm June breezes. Most of the other apples on the farm were red and firm-fleshed Macs growing in a clearing in the woods, but this tree matured lemon yellow fruit with soft white juicy meat. The apples ripened a month before the others. Even while they were still too tart to eat, we used them as missiles to hurl into the frog pond in the cow pasture.

The pleasure of the apple bloom that Robert Frost writes about—the trees' production of sex organs in preparation for a massive orgy—was also ours. But the apple tree is gone, and there are no more crowded orchards on my hill. There are a few stray apple trees growing wild, and then there is the tiny apple bush with the small black fruit, *Pyrus melanocarpa*. Its pink blossoms are smaller than the end of my pencil. In the fall, its wrinkled black fruits (from which the Latin name is derived) look more like shriveled blueberries than apples. *Pyrus melanocarpa* is pollinated by bumblebees, and its fruits are eaten and seeds spread by ruffed grouse, robins, and cedar waxwings.

The color of fruit is important to birds. Fruits become brightly colored when the seeds inside them have matured enough to be dispersed, which is also when they first contain their complement of sugars and attractant flavors. Turning bright red or blue or black, in the case of *Pyrus melanocarpa*, is the plants' equivalent of hanging out a sign that says "food's ready."

To be meaningful, signals should be uniform and unambiguous, and in those of fruit, reds and blues abound. In my forest, the fruit of mountain ash, creeping dogwood, shadbush, pin cherry, choke-cherry, two kinds of holly, two viburnums, hawthorn, and three kinds of lilies . . . are all red. Seven others are blue. Red and blue stand out, and a bird that has identified and eaten one red berry will look for others like it, each in its time. Shadbush caters in June, chokecherries in autumn, and winterberries in winter. We tend to form habits and stop to eat where we have had the best food and the fastest service in the past. As with bees and flowers, the birds that search for berries also need a bright familiar sign telling them there is quick access to high-energy food. They are not looking for novelty and they shun the unfamiliar.

We are perhaps more predisposed to novelty than robins, but we also have favorites. Among most Americans, for example, the more popular varieties of apples are those that are big, red, and shiny, like Red Delicious. Nevertheless, in our taste for novelty we have cul-tured thousands of varieties. All of these varieties only continue to exist through our deliberate cultivation.

We do not know for certain what the original wild apples from which our domestic varieties are derived were like. It was not *Pyrus melanocarpa*. In 1989 U.S. plant scientists tracked down a wild apple species, *Malus sieversii*, in a remote region of Kazakhstan and they suggest that this might be the earth's oldest eating apple. The trees grow along the Silk Route, and traders may have carried its seed throughout the ancient world. This wild-growing apple is still very valuable because it contains genes that make it resistant to many dis-eases and these could be incorporated by plant breeding into modern apple trees, most of which are clones that need constant pampering to stay alive.

As described in a recent article in *Audubon* magazine (November–December 1995) by Mimi Sheraton, the production and preservation of the many modern apple cultivars or varieties is a big enterprise. The largest collection of living apple stock in the world is at the New York State Agricultural Experiment Station in Geneva, New York, operated by Cornell University. The station's nine hundred acres include 3,300 apple varieties that have been developed for centuries all over the world. It is a "gene pool" of inestimable value. It is a Noah's Ark of apple varieties.

Work at the station focuses not only on maintaining a gene "library," but also on developing new apple strains that need fewer chemical fungicides and insecticides, that have longer shelf life, that resist frost damage, that ripen at different times, bruise less, and that have more "eye appeal" to the consumers.

We now guard these specific genetic combinations jealously. My mother has in her garden some apple trees without limbs. They are single-stemmed trees less than five feet tall, with apples growing off to the sides of the trunk for easy picking. They were developed by Stark Brothers Nurseries and Orchards Company in Louisiana, Missouri. Plastic labels attached to the trees warn: "Asexual reproduction using scions, buds, or cuttings, whether for sale or own use, prohibited by U.S. plant patent laws. Infringement will be prosecuted and Stark Brothers offers a $2,500 reward for evidence used in arrest and prosecution of violators."

The genes of apple trees and other organisms are sold and exchanged throughout the world. The semen of a prize bull, or an embryo of a prize cow, may be frozen in liquid nitrogen to be sent by express delivery from one continent to another. It is not necessary to ship the whole bull or calf to get a duplicate or near duplicate. More recently, it has become evident from experiments with sheep that cloning animals from already differentiated body cells is feasible. Similar cloning in plants has long been routine. Most prize apple varieties cannot be reproduced by seeds, because sex (during seed production) destroys their unique genetic combinations. Instead of seeds, buds from a dormant tree in late winter are used for propagating "true to type." They are partially dehydrated to avoid freezing damage, and then stored in liquid nitrogen for up to several years. Upon rethawing they are rehydrated and can then be sent all over the world and grafted onto any convenient apple root stalk, just as prize cow embryos out of a vial can be and often are implanted or grafted to grow in the uterus of any cow. But I digress.

Through the first light snow of December drifting down from gray evening skies, a cow moose and her two nearly full-grown calves walked out of the pines and into the clearing by the cabin. They looked at me with seeming indifference, then came toward the cabin. Unafraid, they walked directly toward the apple tree by the door, and all three of them then started nibbling the tree's twigs. The juxtaposition of moose and apple trees seemed incongruous because one is a symbol of wilderness, and the other a sign of domestication. Yet both are now a part of these woods.

The core of the old tree by the cabin had rotted decades earlier, and there is now an entrance to its hollow heart where a dead branch once broke off. Tree swallows enter this hole and build their feather-lined nests of dry grass in the spring. The tree's trunk has a scar near the base where, in decades past, a porcupine had chewed off a patch of bark. Bacteria and fungi invaded there, rotting the wood and leaving a hole that leads into other chambers of the rotten interior. The

Old apple tree.

thick trunk is twisted and it is pockmarked with small dark spots, the healed sap holes drilled by yellow-bellied sapsuckers in years past before the bark had become flaky and encrusted with cushions of gray-green lichen. Tent caterpillars spin their loose silken cocoons under the strips of bark, after feeding on the fresh leaves in late May. In the past the tree was a part of this hill farm, but now that the farm is gone it has become part of the ecosystem and must fend for itself.

The tree still produces a few apples. By December, the fruit have long since been frost-killed and have shriveled and turned brown. When first ripe they are yellow and have a unique tart flavor that is unlike that of all the other wild apple trees on this hill. They are too tart for me, but they are eaten by bear, deer, squirrels, grouse, porcupine, fox, coyote, and sometimes in harsh winters when pine grosbeaks come this far south they feed on them as well.

I doubt that the tree was planted by the Adamses who once lived on this hill. It is an isolated tree, and it likely grew here from a seed left more casually. Perhaps a red squirrel stored an apple it had found. Perhaps a black bear dropped apple seeds that traversed its gut. Perhaps a child playing in the woods dropped an apple core.

I'd just as soon presume that it came from an apple deliberately tossed from a stick, as I had learned to do from the Adamses for whom this hill was named. We would cut a firm but supple maple sapling and trim it to about five or six feet, and sharpen the tip to a point. With an apple firmly impaled onto the point, we'd swing the whiplike stick sharply forward over our heads. The length of the stick added to our extended arm added force to the end of the flexible whip. When the whip was snapped short the apple would be flicked off, flying through the air until it almost disappeared from sight. Our aim then was the frog pond, far on the other end of the field. The trick to getting it there was to accelerate smoothly up to the maximum speed possible, which of course would be magnified several-fold at the end of the stick, and then to flick the apple off by yanking the switch backward just at the right moment. The momentum would then propel the apple off the stick.

One of the first persons who deliberately spread apple seeds in this country was John Chapman, later called Johnny Appleseed, who was born in Massachusetts. According to legend, he struck out west into the wilderness where he frolicked with a bear, tended a wounded wolf, and talked with the birds. He was treated kindly by the Indians as he wandered barefoot through the Allegheny Mountains, then on to Ohio and Indiana.

He wandered for fifty years and wherever he went he brought apple seeds. He made clearings in the forest and established orchards on the routes he thought the pioneers would take. Johnny died in 1845, after falling ill while trudging through a snowstorm. He had become a legend in his own time.

Johnny Appleseed's apple trees live on thanks largely to the plantings done by animals. In the countryside surrounding Adams Hill, there are few new apple orchards but there are remnants of old ones deep in the woods, and in the woods near my pine grove there are also wild apple trees grown from seeds.

Aside from the apple trees growing from seeds planted by bears, squirrels, and our annual apple toss, I'm not sure whether we're really deliberately planting any new trees at all in Maine, or if we're just propagating old ones that were "born" long ago. Answers depend on whether you look at the question from the practical perspective, or from the perspective of science, or even of philosophy. When it comes to apples and apple trees, the distinction gets wrapped up with "immortality" and "the individual."

Apple trees don't "breed true to seed." All of the McIntosh,

Delicious, and Ben Davis apples in all of the orchards all over the world are ultimately derived from the union of a single sperm and a single egg. They are identical clones—identical to each other yet different from their parents. All have the same DNA fingerprint. There is no more genetic recombination—no more mixing of genes. When allowed to reproduce from their own seeds, genetic recombination "destroys" their type. So they can only be propagated by cuttings. Each variety is a clone. It is as if we could cut off a finger or culture one of our cells and grow from it an identical twin, ad infinitum. Would we ever die? Would we always live on?

Reputedly, in 1836, there were in the garden of the London Horticultural Society already more than fourteen hundred distinct types of cultivated apple clones. On the hill pastures in Maine there are undoubtedly even more varieties than in any register, but none of these are named, because they were not identified and singled out for propagation.

The flowers of most apple trees are much alike and they are not patented, but as Robert Frost and others have long noted, their sudden appearance is startling and beautiful. When the thick buds at the ends of the apple tree's twigs open to reveal five or six flowers, they also quickly unfurl six to seven small, mouse ear–size leaves. The unfurling of the flowers takes precedence over the leaves. Among the flowers on any one bud, the central one surrounded by five others is the first to open. The other five follow so quickly that the tree seems almost to erupt in its blazing show of pink.

A week before the buds opened I had absentmindedly taken a twig and put it in a glass of water inside the cabin by the window. The flowers from this twig opened at the same time as those on the tree, but these were pure *white*, not pink like the others on this tree. I was startled and surprised because I had thought color to be an invariant

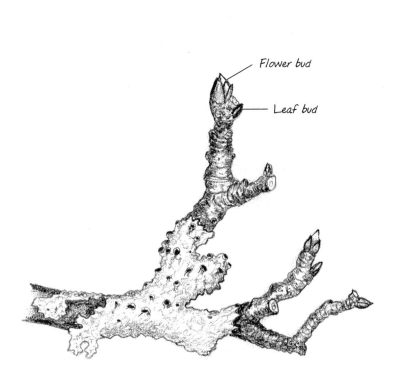

Flower bud

Leaf bud

Twigs from the old apple tree are covered with crustose lichen and represent many years' growth. The young, bare twig is from a vigorously growing young tree and represents only one year's growth.

genetically "determined" characteristic. I suppose it is under "normal" circumstances. However, everything in an organism is determined by its genes, and these genes produce their effects only in the context of specific environments. But being in the glass of water in my cabin window is not "normal"—one more illustration of how genes can produce different effects under different environmental circumstances.

An apple contains five ovaries, each carrying one to two seeds.

Each bud yields one twiglet with flowers and leaves. I counted on average eight twiglets with 6 blossoms on every one-eighth-inch diameter twig. Could such a small twig support forty-eight apples? According to the pomological literature, a medium-size bush tree puts on approximately 3,000 blossom buds, which yield about 18,000 individual flowers. A crop of four bushels of apples, about 160 pounds in most eating varieties, represents the produce of only 600 to 800 individual flowers. Thus, if all 18,000 flowers produced apples the tree would become loaded with nearly 700 pounds of fruit, enough to tear off its branches. This potential catastrophe is always averted, because about 96.5 percent of the fruit is aborted early in development. The small number of maturing fruits relative to what might be produced is not a failure of the tree. It is instead a triumph of adaptation. Trees regulate their own reproduction to produce only what they can support.

After being pollinated, the flowers' pistils start to swell, to become small apples each containing up to ten seed embryos. The tree then begins to shed the surplus apples by loosening the connection of the flower's (young fruit's) stem to the parent twig. The more young apples the tree has, the greater and earlier the abortion schedule. On my tree you can see hundreds of small apples gradually drop off one by one from a single twig, until only one or two remain. These few are then held all the more tightly. They receive the full measure of the parent tree's resources so that they grow into vigorous fruits containing seeds packaged within an offering of food appetizing enough to tempt some animal to eat the fruit and disperse the seeds.

Orchardists whose trees grow under the perhaps unnatural condition of full sunlight (i.e., in an excess of sunlight relative to what they would get inside a forest) help the tree in its abortion process by pruning off a good portion of the bud-laden twigs each fall. Alternatively, they push many of the blossom buds off their spurs while the flowers are still unopened, a procedure called "bud rubbing."

For the lone wild apple tree next to my cabin I do neither, and the moose never browse off enough buds. I simply let the tree take care of the overproduction of seed embryos the natural way. The old tree cannot afford to mature one bushel of apples, much less four. It seems to know how many it can produce year in and year out, which happens to be somewhat less than half a bushel. Yet this old tree's floral display is as vigorous, and as luxuriant, as any young tree's.

A Wild Apple Orchard

It is remarkable how closely the history of the apple is connected with that of man.

—Henry David Thoreau

The remnants of an apple orchard next to my clearing have been long neglected. White pines, red maples, sugar maples, and white ash trees started to grow in the orchard, and during the last twenty to thirty years they have reached up over the apple trees. The apple trees responded by conserving their energy, making fewer blossoms and, of course, maturing many fewer apples. With their meager energy savings, they now send thin limbs probing up to the light, to compete with the other trees. In this they are only partially successful, because the other trees respond in the same way. In this kind of competition commonly called an "arms race," nobody gets ahead, but everybody gets faster, stronger, taller, thicker, or tougher—until some start to lose.

The apple trees in my old orchard are not as good at reaching the light as the native forest trees. They were in imminent danger of being shaded by these natives. The apple crop diminished with each passing year, and I saw fewer deer, bear, ruffed grouse, and porcu-

Recently deceased apple tree in my woods.

pines. Finally I intervened and cut down the largest of the ash and red maple trees, expecting that the apple trees would rebound and again blossom profusely in the spring and produce many apples in the fall.

If I had not come to their aid, all the apple trees would surely have died within ten to fifteen years. I was totally surprised by what actually happened when I tried to rescue them: they all died within just two years. In retrospect I can see why they died. Like many other trees, apple trees long deprived of energy adapt to getting by on less. After having made this adjustment, they are then unable to handle a sudden abundance of light. The sudden exposure to direct sunlight killed them. Birch trees released from shade often react similarly. What is the physiological reason for this death?

As part of the process of photosynthesis plants release oxygen. Too much sunlight may release a toxic excess of oxygen molecules that eventually damages the chloroplasts, the chlorophyll-containing solar cells of the plant. We don't know for sure about apple and birch trees. However, a particularly sun-tolerant desert alga, *Dunaliella*, gains its protection from intense sunlight by synthesizing a protein known as Cbr that binds with a pigment, zeaxanthin, which then diverts the excess sunlight away from the green energy-trapping chlorophyll. Tree leaves have similar pigments. Perhaps these pigments also act like a sunscreen. They are responsible for some of the brilliant fall leaf colors after the chlorophyll, the main energy-trapping compound, is removed from the leaves. Thinking of this I now noticed for the first time that the leaves of white ash trees growing in the sun turned a deep dark purple in the fall, while those in the shaded understory turned a pale bleached yellow, as if they indeed lacked sunscreen pigments. I found the same pattern for arrowwood. The red maple colors, however, seemed wildly idiosyncratic.

Each organism is different, and it is difficult to imagine organisms as varied as algae, apple trees, ash trees, and humans having much in common. Yet generalities of adaptation abound. In retrospect if one might hesitate to give a starving kid a trunkful of candy all at once, perhaps I should also have known better than to suddenly flood my old apple trees in direct sunlight. I might have saved my valued Ben Davis apples from dying prematurely. Sometimes a little knowledge is more dangerous than none.

The Dying and the Dead

There is no death in mortal things, and no end in ruinous
death. There is only mingling and interchange of parts, and it
is this that we call "Nature."
　　　　　　—Empedocles

Some of my oldest and most distinctive trees are sugar maples, two
hundred or more years old. Most of them have remnants of barbed
wire sticking out of them. This wire, their age, and their location
near stone walls are clues indicating that the land was once cleared
and the remaining trees were along field boundaries, where they
then held up fences to retain livestock. Now there is forest all around
and these trees are slowly fading by dropping a huge limb one year,
part of a trunk another. . . . Several of them are between my new
sugar maples and the red spruces on the top of the hill. One in particular has massive lateral limbs and I pause near it often and in all
seasons.

On this day, the twenty-ninth of June, 1996, I've come again to
see what is living at, near, or in this dying sugar maple tree. I've come
with no agenda, and no preconceptions beyond a general familiarity
and attachment. It is a tree I've known for more than twenty years.

The old sugar maple under which I watched.

Another sugar maple.

A mature hemlock with pileated
woodpecker feeding and nest
holes and bees.

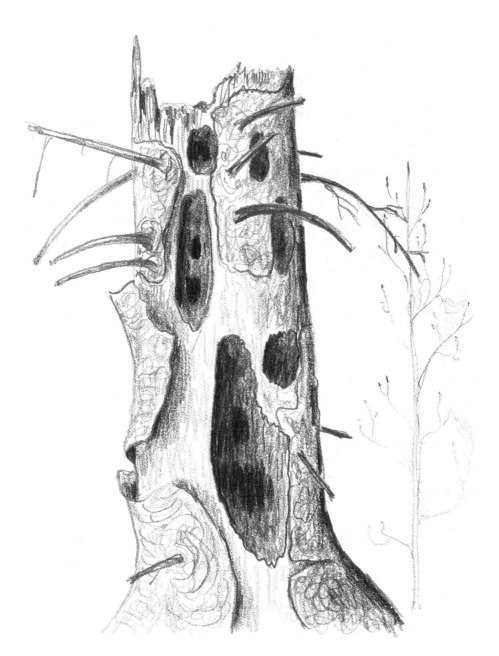

Dead balsam fir, excavated by pileated woodpecker.

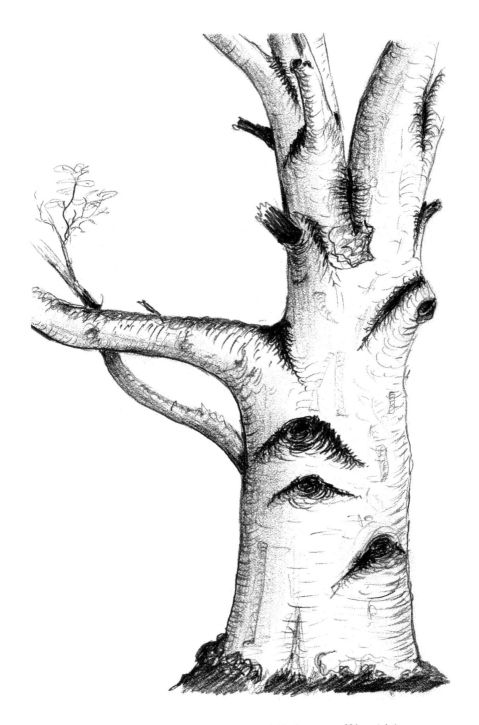

An old white birch that had limbs torn off by old ice storms.

It has been dying all of these years, and it may live, and die, for another fifty years before one May the last buds on the last living twig unfurl. Even then, portions of the tree could remain standing for another twenty years.

More than a hundred years ago the tree held up barbed wire at the edge of a sheep pasture. Branches thicker than my body now reach out to the sides growing almost horizontally from the six-foot-thick trunk. Red squirrels know its hollows and travel on its limbs to the neighboring red spruces. Undoubtedly flying squirrels live here as well, but one rarely sees these large-eyed creatures in the daytime. The tree has a cohort of other, similarly aged maple trees nearby. Some of them have hollow trunks that could become den sites for porcupines or nest holes for barred owls. Perhaps one of them could even be a suitable hibernation site for a black bear. One bear has recently been coming near the cabin eating food I left for ravens, but the location of this tree is too close to my cabin to be prime real estate for bears. They prefer privacy.

Bears were far from my mind as I was perched in semishade among green lady ferns on one of the great limbs that the tree had shed in an ice storm two winters ago. That fallen limb created the light gap where I paused. Within seconds I noticed insects coming and going. What were they, and what brought them, and where do they go? Thoreau in his travels through Maine said: "I am particular to give the names of the settlers and the distances, since every log-hut in these woods is a public house, and such information is of no little consequence to those who may have occasion to travel this way." I also want to give the names of the "settlers"—the trees and insects and birds—perhaps to establish their existence, because their existence is also of no little consequence.

A tiger swallowtail, *Papilio glaucus*, was the first animal that I saw.

It came sailing by, flicking its black-striped golden wings in the sunshine and gracefully maneuvering through a maze of ash, maple, and black cherry leaves. Within ten minutes eight of them passed. In another week, these elegant butterflies will all be dead and not seen again until next year, but they will have left their eggs on black cherry leaves. I know their larvae.

Very young tiger swallowtail caterpillars mimic bird droppings, then they magically shed their skins and metamorphose into smooth-skinned green creatures with two false orange black-pupiled "eyes." They hide on beds of silk in "nests" made by folding the edges of a cherry leaf over and around themselves. Should their disguise and their hiding still fail them, they try their next line of defense. They extrude two bright orange wiggling protrusions from near their false eyes. These protrusions are loaded with chemicals that waft off as a foul-smelling gas. Any bird that has contacted them in hope of a tasty snack will hastily reconsider.

The only caterpillar I saw from my perch on the downed maple limb was a geometrid moth larva dangling on a silk thread from an ash leaf. I presume it had dropped itself from a sapling I had brushed up against on my way here. As it fell from the leaf, it extended a silk safety line from the spinerettes of its mouth. Now that the apparent danger was past, the caterpillar was rolling its lifeline up into a ball with its three pairs of legs, thus hoisting itself back up to the leaf.

Skippers were plentiful. There are several species of these small robust butterflies with earth brown colors and intricate black-and-yellow markings. Many species look similar to me and I can't tell them apart.

That's because I'm not an entomologist. However, as Oliver Wendall Holmes once said, "No [one] can truly be called an entomologist—the subject is too vast for any single human intelligence to grasp." The skippers perched on the ferns close beside me and partially opened their wings. They perched for only a few seconds, then one by one as they warmed up in the sun darted off into the dappled forest. Inexplicably, a swirl of about a dozen of them converged from all sides then flew straight up, spiraling about each other like a miniature tornado, disappearing into the treetops.

Not six feet from my head a blue-black moth fluttered heavily above the young ash trees. I recognized it as the eight-spotted forester moth, *Alypia octamaculata*. It settled briefly with outstretched wings on a leaf and then also disappeared from my view. Moths are stunning creatures. Almost all of the relatives of this moth (family Arctiidae) have gaudy, intricate patterns of pinks, reds, yellows, and metallic blues. Their names—Faithful Beauty, Polka-dot Wasp Moth, Scarlet-bodied Wasp Moth, Snowy Eupseudosoma, Agreeable Tiger Moth, Delicate Cyncnia, Red-tailed Specter, Reversed Haploa, Leconte's Haploa, Yellow Bear, Bella Moth—are as gaudy and fantastic as their colors. This forester with its unique light yellow and immaculate white spots could give a collector heart palpitations, although to collectors *all* insects are beautiful.

Unfortunately, there are not many insect collectors—people whose passion is the beauty and variety of the millions of species. Collecting is the quickest and perhaps only practical way to become familiar enough with these creatures to get to know and appreciate them. What we can't identify doesn't even exist for us. Those to whom all insects are "bugs" feel justified to hang electric "bug zappers" on their back porches. A bug zapper is a commercial device that attracts night-flying insects to a light and then electrocutes them. It is advertised to kill mosquitoes. In one recent study only 31 of 13,789 insects "zapped" at one light over several days were the intended victims. Zappers attract a fantastic variety of moths and other insects but few mosquitoes, and then mostly the males (who don't take blood meals). Crediting zappers for keeping blood-sucking insects away is like claiming that snapping your fingers is what's keeping the bears off your porch.

I had not seen the spotted forester moth for years. A year ago I did see a bluebird in my clearing pluck a number of its brown bristly caterpillars from the freshly growing willow herbs and swallow them whole. It must have been a very hungry bird (a spiny Arctiid caterpillar resembles a bottle brush and when in one's gullet probably also feels like one).

For several minutes a yellow-banded syrphid fly hovered near my head with a high-toned buzz that changed in pitch. The fly rocked back and forth in the air as if suspended on an irregular pendulum swaying in the breeze. A gray robber fly landed on a leaf to my side, cocking its head with bulging eyes in quick jerky motions, apparently searching for prey. Being heated by the sun, it was ready for fast and instant flight. It was hunting other flies. Hundreds of flies of at least a dozen species were everywhere I looked. Most were unobtrusive, but one caught my attention. It flew in with a deep hum like a bumblebee and landed on the log six inches from my leg. A beautiful yellow-furred creature it was, and with its two front legs it proceeded to rub its large multifaceted eyes, then it extended its two front legs, rubbed them together briefly, and flew off in a flash. The larvae of this fly, like thousands of others, is a maggot. This one feeds on decomposing moist wood.

I spied a movement directly next to my leg on the log, a tan brown long-horned beetle (Lepturinae). There are 250 different species of this relatively small group of long-horned beetles in North America. In the spring dozens of different ones crawl and

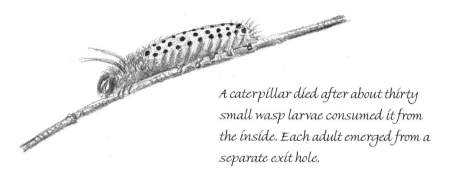

A caterpillar died after about thirty small wasp larvae consumed it from the inside. Each adult emerged from a separate exit hole.

feed in the flowers of the mountain ash nearby, undoubtedly pollinating them. By taking part in the life cycle of these trees they ensure plenty of red berries that feed migrant flocks of cedar and Bohemian waxwings in the fall.

A graceful ichneumon wasp with long, slender, white-ringed legs and antennae hovered around a black cherry sapling. These wasps, like bees, have slender hollow stingers, with which they inject their eggs into caterpillars. The resulting white maggotlike larvae then consume the caterpillars from the inside out. The parasitized caterpillar (with one to thirty or more wasp grubs inside it) keeps on feeding and growing, at first little affected by its terminal "pregnancy."

Another orange-yellow ichneumon wasp appeared briefly. I saw it out of the corner of my eye just as it disappeared into the ferns. There are many hundreds of ichneumon species right here in these woods, and each species parasitizes only specific moths, butterflies, or other insects. In aggregate they are highly effective agents for keeping caterpillar populations from exploding and defoliating the forest.

The cherry tree next to me had been damaged by caterpillar feeding, and the tattered remains of a tent-caterpillar bivouac still remained. Other caterpillars were on the cherry now, and I saw leaf rolls where some of them had hidden. Perhaps these rolls help protect them from some ichneumonids. Despite whatever protection they have, however, they will still not escape these wasps. A few wasp species have managed to search out and reach even beetle and horntail larvae encased in solid wood.

To my right, just a foot and a half away, a small yellow-ringed narrow-waisted potter wasp closely inspected a leaf, then was gone from sight. (The potter wasp builds a small narrow-mouthed pot of clay that after hardening is provisioned with paralyzed caterpillars to feed its growing larvae.) The only insects that stayed near me were those I tried to swat—several deerflies and mosquitoes. Another wasp, a gun-barrel blue spider wasp, landed briefly on my log, flicking its blue wings. It is called a spider wasp because it hunts spiders to feed its larvae. It drugs the spiders to keep them comatose in a secure hiding place, and the larvae then eat them at leisure. Then a long narrow black ichneumon with an ovipositor as long as its entire body hovered near the tree. This one parasitizes horntail wasp larvae that in turn parasitize beetle larvae deep within the solid wood. The wasp needs the long ovipositor to reach the horntail larvae, but it is a mystery how it finds them in the first place.

A half hour passed and I had seen only an infinitely tiny fraction of what there was to be seen. That is the way things are in a healthy forest with many dead trees. So far I've been distracted by the tiny clearing created by just one fallen limb of a dying maple tree. I've hardly had a chance to look at the limb, much less at the tree itself.

The limb that I'm sitting on is covered with a quilt-work of green moss and with chalky green lichens; I also notice patches of crustose bluish green lichens and others that look like white blotches of bird droppings. They fix nitrogen from the atmosphere, making it later available for protein production when it is passed on to plants and animals. I peel the bark, revealing coral-like fans of fungal mycelia, and I wonder what kind of mushrooms they might eventually sprout. I uncover a half dozen or so flat yellow beetle larvae with pincers on their rear ends. Scores of dot-size insects quickly disappear into the cracks. . . . There are many kinds, and I don't have a clue to what they are. This dead limb is the entire world to many organisms, but its influence radiates out into the forest.

Up above me near the recent scar where the limb tore off there are two more large dead branches in addition to one that is still alive and lush with leaves. A hairy woodpecker had drilled a nest cavity into one of the dead branches, and a pileated woodpecker, searching for beetle larvae, has left deep oblong excavations in the other. At the moment some of these holes have threads of spiderweb over them— a sure sign that the hole is "currently unoccupied" by mice, birds, or squirrels. In the fall and winter they will provide refuge for birds in storms and in subzero weather.

On the already dead but still firm wood where the bark has fallen off are tiny round entrance holes to tunnels that the long-horned beetle grubs have made. These are excellent nest sites to about twenty species of wild solitary bees that occur here (principally of the genera *Osmia* and *Megachile*) out of a total of about 3,500 species of other solitary bees in North America. These bees depend on the flowers of the maples, cherries and willows in the spring, and they also pollinate the many berry species that fatten the bears and feed the birds.

The tree also shelters other trees. If the full sunlight were to strike the ground, then a rank tangle of raspberries, blackberries, and grasses could take over. The old tree now excludes these plants and instead the ground is carpeted with a layer of more shade-tolerant sugar maple seedlings. Lack of competitors allows them to take root.

As limbs of the dying tree fall, gradually letting in more light, these saplings will reach higher and begin their struggle to become trees.

The trees' decaying body releases the resources it has collected throughout its life, passing them back into the forest. I do not know whether it is a beginning or an end, or neither. The old tree is surely an island of life in all seasons, and it will likely continue to attract and nourish life for yet another half century in complex ways that I cannot even begin to imagine.

SEEDS AND SEEDLINGS

I'm in awe of this small seed,
Its perfect design and the design within:
The great sequoia tree sleeping, waiting
To send down its first root
No bigger than an eyelash.
                              —Sanora Babb

Most of the trees in my forest face a much more abrupt death than that of the maple tree I watched. They topple in the wind. They are cut down and made into newsprint, computer printout paper, books, and junk mail. Along the path up to my cabin there are not only stumps of cut trees, but there are also wind-tossed quaking aspen with upturned roots. Such upturned roots are almost the only place that the winter wren nests in these woods. When I hear his jubilant song in the spring, I know that an uprooted tree is nearby. Searching for the wren's nests, I usually find it and more.

As I began to crawl under the roof of soil and sod held up by roots I saw a carpet of tiny green flecks—the seedlings of yellow birch. Larger young birch trees were crowded onto the top of the overturned roots. They were taking advantage of direct sun and exposed soil, without which regeneration of this tree species is not possible.

The yellow birch trees with their shiny golden smooth bark are beautiful to see. Their wood is highly valuable. A medium-size tree at current prices would fetch well over five hundred dollars. They are also key organisms in my forest. They provide sweet sap to a woodpecker (see page 173), and the woodpecker is essential for the hummingbirds that live here and that pollinate trees in Central America.

I saw many growth stages of these yellow birch trees where other trees had fallen. In the intermediate stages the dead tree that they grow on rots away, but the birch's roots thicken and become stilts. Eventually the stilts enlarge and merge, extending the trunk down to the ground.

Birch seeds are shed throughout the winter. I've seen them strewn like pepper across the snow for a hundred yards, creating a long thin shadow. They have small "wings" that extend their reach in the wind. The timing of their release on top of the snow surface may also help them travel to find their place in the sun. Initial seedling survival depends on energy supply, either that which they carry along or that which they acquire from sunshine. Birch seeds carry minimal supplies. That is why they must find sun.

In the dicotyledons, the group of plants to which beans, peas, acorns, and beech trees belong, the seed is often greatly enlarged, containing a reserve of food for the embryo provided by the parent tree. The seed is made of two modified leaves, the cotyledons, each engorged with food for the tiny embryo that lies sandwiched in between them. The metabolic machinery of the seed may be shut off for months, or years, depending on the species. For example, beech and red oak seeds are dormant until spring, whereas pin cherry seeds

Blowdown where yellow birch seedlings are starting to grow.

may lie dormant for decades and sprout only when the ground becomes scarred and sunlit. When the seed finds conditions just right, it "wakes up" and begins to metabolize food from its cotyledons, transferring the energy to a rapidly growing root and a stem. The root anchors the seedling and the stem lifts above the ground. The cotyledons may, as in beech trees, also be lifted up, expand, and become flat and green, recognizable as leaves. In other species, such as avocado and oak, the cotyledons remain under the ground.

There is amazing variation in the amount of food different trees provide to their offspring for their start in life. A tree has only so much to give, and it can have either millions of seeds each with a tiny inheritance, like a birch, or relatively few seeds each endowed with

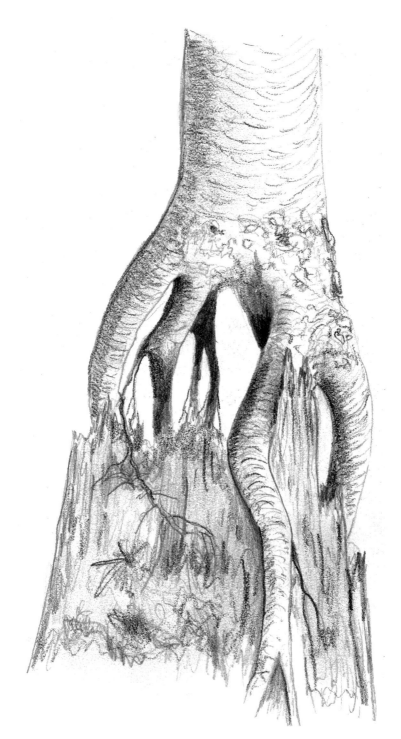

Yellow birch "stilting" on a gradually decaying stump.

a large inheritance, like an oak. In my forest the poplars are an even more extreme example of the first strategy than birch. In May quaking aspen and big-tooth poplar seeds float gently through the air on white silky parasols to settle on the ground in untold millions. The ground everywhere becomes covered with a white downy film. The seeds themselves are almost invisible flecks. I weighed several, and along with their parasols, they weighed barely 0.1 milligram each. Being without food reserves, these seeds must have sunlight immediately after sprouting in order to remain alive. They must find the very rare patches of unshaded uninhabited land, like those occurring after a fire has recently cleared the trees. As such, they are "fugitive" species that may sprout far from home. They need to produce many seeds all the time so at least some stand a chance of being in the right place at the *right* time. Hence all of the tree's apparent excess seeds are not excessive at all.

An aspen seed's chance of growing into a tree is astronomically small. Once an aspen tree has established itself, it propagates vegetatively by sending out runners under the ground. The eastern slope of my hill was covered with an almost continual stand of thousands of quaking aspen. It is likely that most of the trees were one or several genetically identical clones originating from one or several seeds. Now, after most of them have matured and have been wind-thrown or cut down, their roots are again sending up a thicket of genetically identical shoots. The shoots are all growing at nearly five feet per year. In twenty years, these genetically identical individuals will again be standing as before and they will send out billions of seeds on the wind every spring.

Acorns and beeches are, in comparison to aspen or poplar seeds, gigantic. With their large inheritance of food they can grow for a short while even under the shade of the forest canopy. This

The two cotyledons and first two
leaves of seedling maples (left) and
beech (right).

competitive edge allows them to get started even within the mature forest.

One problem with loading up the seeds with a large food store is that it invites predators to eat them. Another is that mobility can be low. To some extent, the seed predators have been exploited by oaks and beech trees to counteract the second problem. Blue jays and squirrels eat nuts, carrying off and planting the surplus they can't immediately eat. Most other seed predators are not so helpful. To have a surplus of seeds that will not be eaten, the tree generally produces *no* seeds for a few years—until all of its seed predators have starved or left. Then the tree suddenly puts out a giant crop that hopefully won't be completely devoured.

I have kept informal records of the nut crops, commonly called "mast," in my forest since the fall of 1980. In 1980 there was an incredible and highly conspicuous crop of beechnuts, acorns, and sugar maple seeds. On the southwestern slopes of Gammon Ridge near my cabin, there was constant activity of black bear and deer, as well as chipmunks, red squirrels, and gray squirrels. Flocks of evening grosbeaks were feeding on beechnuts, and blue jays took part in the harvest. I even saw a hairy woodpecker, a chickadee, and a red-breasted nuthatch feeding on beechnuts. Insects were starting to zero in on the crop as well. Twenty-four of 303 nuts I examined in October were parasitized by insects. By November 14, most of the remaining red oak acorns were

Aspen seed

Basswood

Seeds of aspen and basswood are dispersed by "parachute" (left) and those of beech and oak by jays and squirrels.

also parasitized, each by a little white grub, a beetle larva. I do not know how many of these were simply the rejects that the birds and mammals left after taking all the rest.

The next year there was no mast at all. During fifteen years when I kept informal (unfortunately incomplete) records, I found that in seven years the beeches did not have nuts. In six years the beechnut and acorn crops coincided in the same year, but in two other years acorn and beechnut crops were large, but in separate years. There was no way of knowing how many seeds were left over in this forest for regeneration, or in which years regeneration occurred.

Sugar maples, like oaks and beeches, also produce seed crops only in some years. Their "pulsing" of seed production results in even-aged cohorts of young. In a mature sugar maple woodland with no underbrush I found a small hole in the canopy where a tree had fallen. Sunlight had penetrated, and the ground was almost a solid mat of sugar maple seedlings, averaging four to five inches tall. I pulled up seventy-one of these seedlings from a four-square-foot plot. (In nearby areas under more solid canopy cover there were no seedlings at all.) This density of nearly eighteen seedlings per square foot covered an area of some five hundred feet, and, in total, the "hole" where the tree had fallen must have contained more than ten thousand seedlings. Only one of them would survive to replace the

Airborne acorn disperser

blowdown tree and eventually produce seeds of its own. I could pull up seedlings here all day and scarcely have any impact. Even as I was pulling up my samples to look at them more closely, I noted that many of them were dead already.

The seedlings were tiny. Their diameter varied from 1.5 to 2.5 millimeters. Were they all from one year's seed crop, perhaps last year's? To find out I cut their thin stems with a sharp razor blade and counted their annual growth rings under the microscope. To my great surprise, in the sample of thirty-five that I examined, the growth rings indicated ages of six to ten years. The most common age was seven, and deviations from that were within my error of counting, given the difficulty of seeing, much less counting, the microscopic growth rings. Apparently there is not a new crop of seedlings every year. Instead, in those few places that seedlings find sufficient light to sprout at all, they barely hang on for many years as a light-hungry crowd. Gradually they die. The parent tree, for all these years, has little need to produce more seeds, because those it has already produced still lie in waiting. To the parent, it makes no difference which of its many offspring will be the lucky one that inherits its space. By generating a crowd of seedlings under it, might a tree effectively reserve the space so that competing trees cannot also have offspring-in-waiting at the same spot? How many maple seeds does it take to make one seedling?

In 1982, when I first wondered, I took on a summer project with my former wife Margaret to find out. We marked eight four-square-foot plots in each of three different areas. We removed and counted all the existing maple seeds (all red maple). There were 57 to 121 seeds already present per plot (no seedlings present). We removed the existing seeds and then added 2, 10, 50, 75, 100, 150, 250, or 500 red maple seeds of our own that we picked from adjacent ground. In 1995, thirteen years later, *none* of the plots had maple seedlings! Did rodents eat all the seeds? Did the parents' shade inhibit all new growth?

The only additional step, beyond leaving a crowd of seedlings, that the parent tree could take to help fight off competitors would be to commit suicide. Robin S. Foster of the University of Chicago reported that *Tachigalia versicolor* from lowland forests of Central America does just that. It is a large-seeded, canopy-forming tree with a large crown that apparently only flowers once in its life. Then it dies and sheds its seeds, which are rarely found more than 100 meters from the parent. Stem growth rings of this tropical tree are not distinct due to the constant growing conditions, so that the tree's age at reproduction and death is not known. However, trees with a trunk diameter of 116 centimeters and a height of 35 meters can't grow overnight. For a tree to delay reproduction for a one-shot deal after reaching a venerable age and height well above the canopy is unique. Foster concludes: "A suicidal act of reproduction might seem maladaptive for a species of huge trees. The trees do not seem to store nutrients and energy for the one reproductive event, for the quantities of flowers and fruits seem no greater than in comparably-sized tree species with repeated reproduction. On the basis of initial observations, I propose that selection has favored individuals which, by dying and falling, greatly increase the probabilities for their few progeny of survival to maturity in the forest canopy. The parents' death creates a more favorable environment for the development of its offspring, and occurs when the offspring are still abundant—having not yet succumbed to all other sources of juvenile mortality."

Suicide would not be favored where seeds can be widely dispersed to find openings in the canopy. Temperate trees like those of my forest also may not have the same constraints as tropical trees that seldom experience large light gaps, and where suicide is needed to create an opening. However, when a temperate tree is weakened, it might be advantageous for it to gamble and reproduce one last time before dying. This would have the appearance of suicide. Maybe it is suicide.

ACORNS

The ecologization of politics requires us to acknowledge the priority of human values and make ecology part of education at an early age, molding a new, modern approach to nature and, at the same time, giving back to man a sense of being part of nature. No moral improvement of society is possible without that.

　　　　　　　　—Mikhail Gorbachev

The red oaks in my forest shed most of their acorns in the last week of September, when the fall foliage colors first appear and just before the leaves fall. The timing is probably right, because the dead leaves bury the seeds and potentially hide them from some predators, while still leaving many available for the seed dispersers.

The weather is usually gorgeous at this time of year. It is hard to stay indoors and excuses to be outside come easy. Why not a family project like picking acorns? My thirteen-year-old son didn't think it was a good idea. He thought it was definitely uncool. But we did it anyway. As it turned out, I picked most of the acorns, perhaps because an adult mind can reach further into the future. . . . I was envisioning oak trees, mast, turkeys, and gray squirrels. . . .

Stooping to pick acorns off the ground would seem to be the most harmless and innocuous thing to do in the world. Or was it the most potent? The first time I stooped to pick up acorns where they were most conspicuously exposed, along the sides of a gravelly road, a man came by in a pickup truck. He slowed way down, watching me warily through his rearview mirror, as if I were involved in some kind of subversive activity. He went about fifty yards past me and then came to a complete stop. Fifty yards was as close as he dared to come, because as soon as I started to walk toward him he sped away.

I picked up more than a bushel of red oak acorns. There was amazing variation in nut shape, color, and size. Whole nuts ranged in weight from 0.85 gram to 7.47 grams (those unparasitized by the snout beetle larvae, which leave a small round hole). While some individual trees produced uniformly large round acorns, others seemed to produce only small round or oblong ones. To confirm my impression of variety in fruit between trees, I kept samples of twenty acorns from six separate trees in six separate bags, weighing each acorn and then calculating the average weight. Under one tree, for example, where the ground had been strewn with thousands of acorns, the average acorn weight was 2.60 grams. Under another nearby tree, the average weight was 5.95 grams. Variation is a substrate for natural selection to work on, if it turns out that conditions change and one or another size becomes more adaptive. If, for example, gray squirrels disperse and bury the smaller acorns more than the larger nuts, then trees growing where the squirrels are abundant will reproduce more if they have small nuts. Perhaps, if a very large disperser were available then acorns would grow to be huge. A botanist friend of mine who takes his class on a field trip to Costa Rica every year impresses his students with the coffee

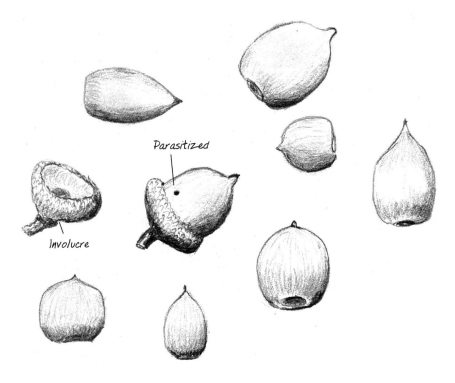

Parasitized

Involucre

Average shape and size of acorns from eight different red oaks growing within two hundred yards of one another.

mug–size acorns that can be found there, and his students go bug-eyed when he says, "And you ought to see the size of the squirrels!" (The acorn dispersers there may be monkeys.) Alternately, if survival of seedlings depends more on the seedling's being able to compete or grow for a time in shade, then trees with larger acorns should be favored.

I picked the acorns mostly to plant in my pine grove that I had just thinned out. It was a cow pasture within recent memory. I wanted to increase its diversity, to make it a forest. It is a truism that in an ecosystem everything is connected to everything else. This often makes predictions difficult, because almost anything one may

care to examine can affect thousands
of others down the line, and each
of the cascading effects may
proceed at different rates.

One thing the oaks will add is
greater insect diversity. There might be
an increase in hundreds of species of
moths, beetles, flies, and their ichneumon wasp parasites. There might
also be more species of fungi.
The bird fauna might enlarge
to include more individuals and
species of warblers, woodpeckers, and flycatchers.

The largest effect might occur after
the trees start to produce their own
acorns. From previous studies we know
that acorn crops lead to population
increases of deer mice and presumably other rodents, such as chipmunks
and flying, red, and gray squirrels.
Certainly there would be an influx of
white-tailed deer feasting in autumn on
acorns. Wild turkeys might even be
attracted. I've seen a flock thirty miles
south of here.

The cascading chain of events and
interactions among these animals is

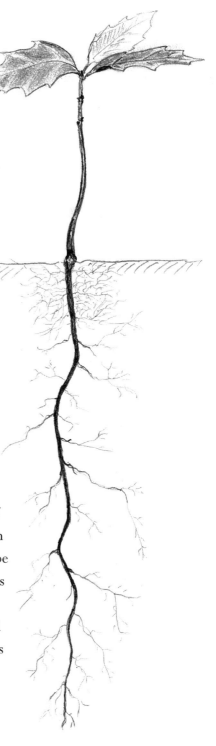

An oak seedling.

unpredictable and apt to be surprising. For example, a severe invasion of gypsy moths has the potential to denude the forest, killing many trees. Experiments have shown that deer mice can prevent outbreaks of these insects, because these tree-climbing mice avidly hunt for the moth pupae in winter and they can nearly eliminate the moths from a forest. On the other hand, if the moth populations increase unchecked and do kill trees, then the underbrush will grow and other tree species can take advantage of the space. Defoliation by gypsy moths reduces or prevents the oaks from producing acorns, thus reducing the rodent, deer, and turkey populations. The whole system could then balance on a virus affecting the moth.

After I had picked sufficient acorns to plant in my woods where many animals would eat them, I decided to try some myself. Native Americans used to live off acorns, and people ate them in Europe during the hard times after the last war. However, some acorns taste much better than others. There are about eighty-five species of oaks in the United States, and they are divided into two major groups, the white and red oaks. All acorns can be eaten, but those of white oaks have fewer tannins than the reds and they are preferred. Tannins are bitter-tasting chemicals with which plants protect themselves from predators, much as a skunk protects itself by its stink. Tannins, soluble in water, are not harmful unless ingested in large amounts. I read that acorns "can be made sweet without great effort." The method of sweetening acorns is similar in principle to that used to de-scent a victim of a skunk attack, only a little more radical. The Native Americans of various tribes leached bitter acorns by methods that included (1) burying them in the mud of a swamp for a year, (2) burying them in fresh water under sand, and (3) pounding them into meal then allowing fresh water of a stream to run through it for at least a day. The recipe I had said, "It is an easy matter to leach bitter acorns at home today. Just shell the nuts and boil them whole in a kettle of water, changing the fluid when it yellows."

The raw red acorns without treatment indeed tasted vile, but the acorns that I gathered from under a white oak on the University of Vermont campus were only mildly bitter. These might be the ones I should, as a neophyte of acorn cookery, start with. I'd boil them, then roast them, and then use them in bread, cakes, pancakes. . . .

My shelled acorns were gleaming white. Within minutes, the boiling water I dropped them in yellowed, then turned dark brown. I replaced the water and boiled them again. The water quickly turned dark brown again. I repeated this eight times. By the time the acorn meat itself had turned brown I decided to leave well enough alone and passed boiled acorns all around for my son, my wife, and visiting friends to try. Reactions at first bite ranged from "mm, not too bad," to running out of the house and spitting wildly into the bushes. As for myself, I think I could eat them, if I were very hungry.

I planted my surplus, unboiled acorns (and along with them a smattering of hickory nuts, black walnuts, and beechnuts) by pressing them into the soil in light gaps where I'd just thinned out my pines. Given the fresh light gaps, some seedlings might have a chance. Since it will take perhaps a hundred years before a few of these seeds will become trees reaching into the canopy, *now* is the time to start, not later. Now is the time to give a few nut trees the chance to help re-create the diverse forest that must have been here a long time ago.

Planting nuts requires a vision for a future that goes beyond one's mortal reach. If we envision ourselves as participants in the same grand, complex web of interactions as the forest, then planting acorns is like planting part of ourselves. The morality that comes from such a vision of ecosystem-as-life is a common thread that, if taught and encouraged, could unite all of mankind.

OF BIRDS, TREES, AND FUNGI

When we try to pick out anything by itself, we find it hitched to everything else in the universe.
>—John Muir

A patch of mature quaking aspens along the side of the path to the cabin is slowly being overtaken by maples, beeches, and white birches. The aspens of this patch are all still alive and did not yet seem suitable woodpecker nest sites. However, I noticed yellow live wood showing just beneath the bark on one of these trees around a woodpecker hole about twenty feet up. This hole must have been freshly excavated. I then saw the owner pop out of the hole—the yellow-bellied sapsucker. This species, unlike most woodpeckers, does not generally drill into hard wood for food.

There were three fruiting bodies of a polypore fungus (later identified as *Fomes igniarius* var. *populinus*) near the fresh woodpecker hole. Most polypores grow on dead trees. The woodpecker had probably chosen this tree because the wood was being softened by the fungal mycelia.

As I later learned, this was a typical sapsucker nest site. On May

Sapsucker next to nest hole on live aspen with two Fomes igniarius fungal fruiting heads.

22, when I first glimpsed the nest hole, the female bird that came out perched next to the hole, and the fungi. With jerky movements of her head, she alternately looked into the hole and toward me. Then she flew off on rapid wingbeats into the surrounding forest, and as I was about to continue up the path, I heard a muffled hum next to my ear. A male hummingbird, his ruby throat flashing like a jewel, hung briefly in the air by my face, then he sped off, flying low over the forest floor under the unfurling pea green beech leaves. Was the presence of the hummingbird, like that of the fungus, also no coincidence? Was there some connection among the sapsucker nest site, the fungus, and the hummingbird?

There were no flowers nearby to feed a hummingbird. This northern forest seems an unlikely habitat for a bird just returned from Central America where it spends nine months of the year feeding on nectar from many red tubular flowers. Red has special meaning to hummingbirds. Even here in the northern forest, hummingbirds fly to and investigate red objects, even people with red in their clothing. There are no red flowers in these woods; in fact, there are hardly any flowers at all this early in the spring except a few unobtrusive white ones on the ground.

Here the four-gram ruby-throated hummingbird satisfies its voracious appetite for high-energy food on an alternative energy source: the sweet sap issuing from the holes made by the sapsucker in birch trees. Since sap does not run at subzero temperatures sapsuckers are also migratory, unlike the other three local north woods woodpeckers. The rubythroat times its spring arrival into these woods with the arrival of the yellow-bellied sapsucker so it can take advantage of his sap founts.

One of the first things a sapsucker does after returning in the spring is to find a vigorously growing birch tree. It chisels off bark in

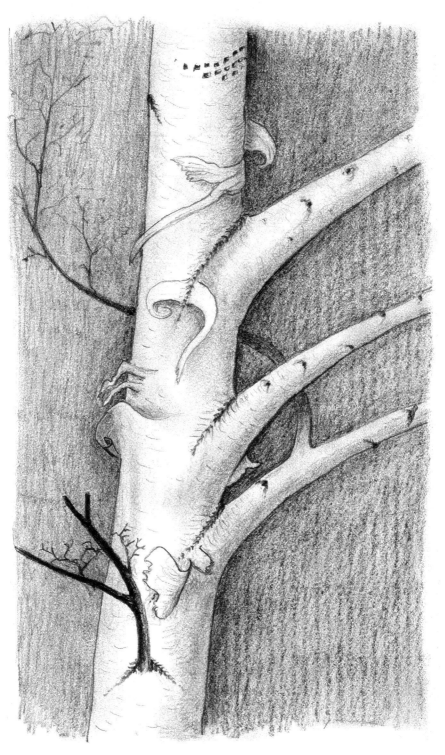

A white birch with a beginning yellow-bellied sapsucker lick
(near the top of the sketch).

a pattern of little squares. It removes the outer bark as well as the inner bark (the phloem) down to the surface of the wood (xylem). A series of sap wells are made side by side, girdling the whole tree. Subsequent rows of wells are made above previous rows, often those from the previous year. I have tried to make similar wells, but I failed in generating sap from them. The sapsuckers lick with their long sharp pointed tongues at the upper ends of the wells to keep sap flowing, perhaps repeatedly injuring the phloem sieve tubes that the tree normally "heals" to shut off sap loss. I had neglected to do the same.

When the hummingbird approached me (probably because I wore red), I thought of its dependence on the woodpecker I was watching. I wondered how the hummingbird might locate sap licks, since sap licks, unlike the flowers that the bird visits, are difficult to see, nor are they red. How could the bird suddenly switch from looking for bright red conspicuous flowers to unobtrusive little holes in tree stems? Why would a hummingbird seek food at such an unlikely source? In this vast forest there are thousands of birch trees and most have no sap licks. Could the bird be attracted initially to the bright red coloring on the woodpecker itself as it tended its sap wells? Sapsuckers spend a lot of time at their sap licks, and both the female and the male have red coloring. Every day I saw one or a pair of them at "their" white birch near the edge of my clearing, along with at least two hummingbirds (and numerous white-faced hornet queens).

My casual observations provided no insights, just questions. I saved myself thousands of hours of potential work, yet found answers to some of these questions in just a couple of hours by reclining with reprints in one hand, and a cup of coffee in the other. There are times when it makes sense to be an armchair biologist, especially when one bumps into hundreds of questions and has only one lifetime in which to seek answers. The scientific publications I read were by Paul R. Ehrlich, Gretchen C. Daily, and Nick M. Haddad of Stanford University. They worked for many summers at the Rocky Mountain Biological Laboratory (RMBL) near Crested Butte, Colorado. Paul Ehrlich, famed among biologists for his work on butterflies, and known to the general public since his book *The Population Bomb*, watches sapsuckers for fun "more than any other bird."

The Stanford group studied the red-naped sapsucker (*Sphyrapicus nuchalis*), a close relative of the yellow-bellied sapsucker (*S. varius*). In the subalpine ecosystem near the RMBL the red-naped sapsucker creates extensive sap wells in willows rather than birch. Just as in Maine, the sapsuckers there also drill their nest holes each year in live quaking aspen, and also only those infected with the same heart-rot fungus. The study group determined that the woodpeckers' distribution in Colorado depends on this fungus, on aspen, and on willow thickets. In Maine it probably also depends on aspen and this fungus, and on birch trees instead of willow?

The sapsuckers, it turns out, affect the ecosystem in a variety of ways. First, the nest holes they chisel are subsequently used by other animals. In Colorado the sapsuckers provide at least ten times as many nest holes as any of the other, much rarer, woodpeckers. The nest holes they make are essential for tree swallows, violet-green swallows, house wrens, mountain bluebirds, mountain chickadees, and other birds.

The impact of the woodpeckers on the populations of the hole-nesting birds can be evaluated, but the subsequent impact of all those sapsucker hole-nesting birds on the rest of the ecosystem is "unknowable," though presumably great. For example, all the hole-nesting birds are voracious insect predators, ultimately protecting aspens, willows, and other trees from defoliation. The humming-birds, who also rely on the work of the sapsuckers, are pollinators necessary for plant reproduction, although the rubythroat is a polli-nator in another ecosystem a few thousand miles from my forest. In Colorado sapsucker wells provide abundant nourishment (sugary syrup) for at least two species of hummingbirds, for orange-crowned warblers, tachinid flies (important insect parasitoids), vespid wasps, and occasional chipmunks and red squirrels.

I suspect that the role of the hornets and other vespine wasps that visit the sapsucker's sap wells in spring is particularly impor-tant. Vespine wasps are predators that catch and consume other insects. In 81 percent of the Stanford group's observations, one or two *dozen* of the wasps were simultaneously present at each well, where they secured sugar water to power their hunting forays. The close interrelationship of the numerous participants in this ecosys-tem, it was concluded, shows that the disappearance of any one ele-ment of the complex "could cause an unanticipated unraveling of the community."

I have no idea what the ultimate effect of sapsuckers are on the Maine woods. However, these birds are surely one of the innumer-able, but generally invisible, links in the complex forest ecosystem. It is safe to conclude that sapsuckers are a "keystone" species, analo-gous to the keystone in an archway. Removing the keystone of an archway causes the whole arch to collapse just as removing a key-stone species from an ecosystem creates a cascade of effects. In the complex construct of an ecosystem, one could just as well call the aspen, or even the heart-rot fungus, or the willow/birch, the key-stone organism. All are essential ingredients in the complex interde-pendence that has established itself over millions of years.

MUSHROOMS

In nature's infinite book of secrecy
A little I can read.
    —William Shakespeare

Ten years ago, I had chopped down the brushy pines in the over-grown old apple orchard just north of the cabin, creating light gaps. Sugar maples have extended their branches into the light gaps and filled them. The ground here is illuminated through layers of leaves in soft luminescent green, and all around red-eyed vireos sing for most of the day. Since direct sunlight no longer reaches the ground in the summer, the ground stays moist. It was especially moist in the rainy summer of 1996, making conditions ideal for mushrooms.

It was June when I first noticed the mushrooms—a group of bright orange caps on slender yellow stems. Within days scores of them had erupted. Throughout the whole maple grove there were patches of brilliant orange caps on the damp brown earth. Did these fungi, like others, form symbiotic relationships with the roots of the maples, enabling them to better extract nutrients from the soil?

Most of the body of a mushroom-forming fungus is composed of long thin filaments, or hyphae, that reach underground in all direc-

Lemon yellow

Brown

Mushrooms in July in the maple grove.

tions. These hyphae are invisible to a casual observer like myself, but they are the "working" part of the organism. They could have been growing underground among the maple roots for years. The bright fruiting bodies we call mushrooms are only temporary structures for making and dispersing millions of microscopic reproductive spores.

Mushrooms that erupt from the ground in groups, sometimes rings, are likely part of the same organism that is connected underground. The hyphae of some species grow for countless years. One toadstool fungus has been reported in Michigan that is of startling dimensions. Genetic evidence indicates that the same individual (or parts of it?) stretches over hundreds of acres. It has likely consumed generations of dead trees within its realm, converting them into humus and releasing the nutrients the tree had accumulated over a lifetime. These nutrients make the growth of new trees possible.

Fungal hyphae not only grow and consume dead trees, recycling them back into the ecosystem so that new trees can grow, but also grow in a very intimate relationship with, and on the roots of, living trees, forming a symbiotic association called *mycorrhiza* ("fungus root"). These fungi need the living trees for their energy resources. Indeed, trees have even been called "the photosynthetic appendages of fungi."

While the fungi depend on the tree to get carbon, the mycorrhizae in turn help the tree's roots to extract hundreds and up to a thousand times more phosphorus, nitrogen, zinc, copper, manganese, and other essential nutrients and water than would be possible without them. The teaming up of fungi with plants to produce the mycorrhizae is a powerful boost to both that made it possible for some vascular plants to evolve to large size. Some biologists even suggest that it may have been a decisive step that made the evolution of big trees possible.

Different species of trees are often associated with their own unique species of fungi, although some mycorrhizal fungi colonize several species of trees. We tend not to notice these fungi until they send up their distinctively shaped and colorful fruiting bodies that are good to eat, such as truffles. One group of about two hundred species of mycorrhizal fungi known as truffles has been prized by human (and other animal) gourmets for centuries. Each truffle has its own species of tree host. The black Périgord truffle, for example, grows on the roots of oaks and hazelnuts. Animals (such as pigs) locate truffles by their powerful scent, and when excavating and eating them the animals disperse the fungi's reproductive spores. Wild boars are thus also important components of some ecosystems.

Our understanding of mycorrhizae began with truffles. Back in 1885, the king of Prussia commissioned Professor A. G. Frank, a forest pathologist of the Landwirtschaftliche Hochschule in Berlin, to try to grow truffles. In the process of trying to grow truffles, Frank unlocked the fungi's connection to trees. He determined that the fungus acted like "die Amme" (the wet nurse) that literally nourished the tree. Further, he found that the association, which he called "mykorhizae" from the Greek fungus-root, was widespread, existing in all of the individual trees of the several European species of trees he examined. He revolutionized our understanding of the importance of symbiosis in general, and he showed us the ecological importance of fungi to forests specifically. In a recent book on mycorrhizae, Michael F. Allen duly acknowledges that Frank failed with his original objective of growing truffles, but he concludes: "We all should fail so nobly."

Once "tuned in" to mushrooms, I began to see them everywhere all summer and on until late fall. Even after the leaves were down, there were still periodic "blooms" of different mushrooms. The forest floor seemed like the skin of a chameleon with patches of continually changing colors. There were patches of brown, lemon yellow, purple, red, orange, and tan. There were even green and white mushrooms. I tried to name some of them, but rarely succeeded. Many are good to eat. Many are poisonous. I was never confident enough of my identification to risk my life and eat them. My four different mushroom books showed hundreds of species, but each book showed different ones, and even pictures of the same type looked quite different. My wife, Rachel, kept finding ever-more species, and she *did* identify a dozen or so. I even trusted her enough to eat one that she recognized as Dryad's Saddle, *Polyporus squamosus*, from a dead elm, declaring it to be edible. Well, so were acorns. Maybe it's the cooking? In this case it wasn't the cooking. I fried it myself, and it truly *was* delicious!

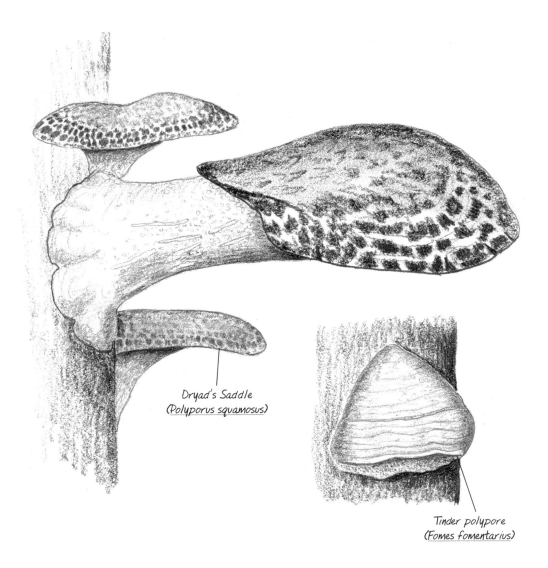

Dryad's Saddle
(*Polyporus squamosus*)

Tinder polypore
(*Fomes fomentarius*)

I'd eaten mushrooms before. As a child, my mother took us into the forest to collect them for food. I learned to recognize the beautiful but poisonous red-capped and white-speckled amanitas. I learned a certain poisonous bolete blushed out in purple when you bruised it. We ate six to eight species. They included brown boletes with yellow and some with white pores as well as chanterelles.

The fruiting body of a polypore fungus on a living beech. The beech is growing around the intrusion.

Several species were especially delicious, and there was great excitement when we found them. Later, our family became more prosperous. We stopped collecting mushrooms. The fungi lost their immediately perceived importance to me, and much of their immediate fascination as well. That is the way of things. Young humans are much like other young animals. They pay most acute attention to matters of the stomach.

I did not now dare to indulge in eating boletes, my wife's growing expertise in identifying them notwithstanding. In any case, most of the potentially edible mushrooms disappeared by the end of September. Individuals of many species last only several days before dissolving into a black ooze. The many (mostly inedible) species that grow on trees and other deadwood remained. Most of those on trees are of a woody or leathery consistency, and they appear to be nearly indestructible. Some stay on a tree, growing for many years, showing easily recognizable annual growth rings, like a tree's wood. Sometimes they stay long enough that, when they invade live trees, which is rare, the trees' wood grows around them.

In November I was again immersed in mushrooms. The abundance of bracket or shelf fungi (polypores) on recently fallen trees and stumps had been there all along. I just hadn't noticed. But after one day of searching in the woods, I had found twenty-one species. After the second day the number was up to twenty-five. Perhaps this was not a bad "catch." However, it is not as good as it gets. There are an estimated million species of fungi, but only about sixty-nine thousand are named. Undoubtedly, all of the polypores I found *were* named, but I could name only a few of them. I sketched the whole lot so that I could at least recognize them again in the future and count them among my acquaintances. Those whose names I did not know got nicknames. I always remember a "face," seldom a name.

I soon noticed more than the "faces." I also noticed that on any one tree, or on any one area of a tree, only one type of mushroom grew in great numbers, and all about the same age. Other types would be absent in that spot, even though they were common nearby. It seemed as though after a dead or dying tree got inoculated, the fungus "body" spread quickly throughout the tree. The fruiting "faces" are merely symptomatic of a much greater fungus "body." Why did any one tree, or patch, not show evidence of many species of fungi? Do fungi wage chemical warfare to exclude each other? Do they fight each other, and bacteria, with antibiotics? It would be surprising if it were otherwise. After all, penicillin is precisely this—a fungal antibacterial weapon, from the fungi we now call *Penicillium notatum* and *P. chrysogenum*.

Not one square millimeter of my forest is fungus free. Every breath we breathe is populated with fungal spores. Every single tiny mushroom growing in damp decaying leaves, and every polypore on a dead tree, disperses spores by the millions. The ground and deadwood are riddled with networks of fungal hyphae. Some even grow invisibly in living trees.

Recently Leo Pehl and Heinz Butin from the Forestry School in Braunschweig, Germany, have started to uncover a new group of fungi they called "endophytes" (with as many as one hundred different species colonizing any one tree type) living symptomless and unseen as friendly tenants in trees. These fungi become apparent only through their action, which is to prevent destructive fungi and also animal pests from harming the tree. In one example that Pehl and Butin documented, an endophytic fungus living in oak leaves acted analagously to a mine that is harmless unless stepped on. In this case, the "mine" does not explode until a midge lodges in the leaf and causes it to grow a gall. That is, the midge infection activates the fungus, and it then assists the oak leaf in getting rid of the parasitic midge, in effect "paying" the tree for its residency in it.

We have an irrational fungi-phobia that costs us billions of dollars, and it robs us of riches and valuable time. It leads us to damage our health by using huge amounts of synthetic fungicides and pesticides. Apparently we fear some fungi even more than the toxins that we use to kill them. Fungal apple scab, for example, is one feared species, though it really does only minor damage. Scab costs Vermont apple growers alone one million dollars annually in chemical applications and "product loss." (Scabbed apples are "unsellable.") Yet, a little apple scab is harmless and tasteless. Given the choice, I'd purposely *pick out* apples with *some* black fungus scabs, because they could not more honestly be labeled "fungicide free," in the same way that a tiny tasteless moth caterpillar in the apple core says "insecticide free."

ANTS AND TREES

Insects won't inherit the earth—
they already own it now.
                    —Thomas Eisner

I can personally attest to the effectiveness of ants as tree guardians. Stuart and I had built a tree house out of old boards in the large red spruce growing in our clearing by the cabin. It is pretty fancy for a tree house, having a floor, roof, and sides made of scrap lumber left over from building the outhouse. Soon after we built it we noticed the ants, *Formica subintegra*. This big ant with black abdomen, dark red-brown head, and thorax is equipped with a pair of strong sharp pincers. They were marching up the trunk of our tree in columns, then dispersing near our tree house up in the spreading branches of the tree's crown. Some invaded the tree house itself. We put grease around the trunk of the tree near the ground, but they somehow managed to breach that sticky moat. Stuart finally abandoned the tree to them. I suspect small caterpillars that eat leaves would do no less. That is, the ants protect the tree from insect predators.

Insects are ever-poised to devour living trees. At times they succeed, but trees and other plants defend themselves with a variety of

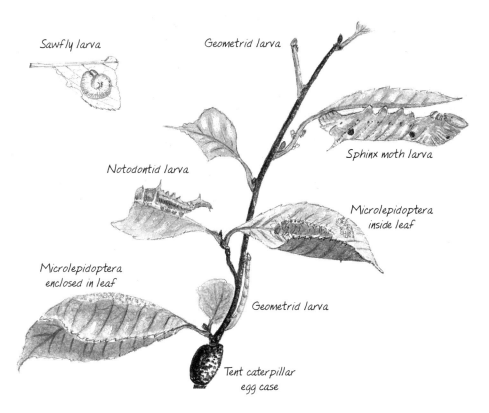

Sawfly larva

Geometrid larva

Sphinx moth larva

Notodontid larva

Microlepidoptera
inside leaf

Microlepidoptera
enclosed in leaf

Geometrid larva

Tent caterpillar
egg case

A sampling of insect herbivores on a small chokecherry tree behind the cabin. Three aphids are shown, in addition to a tent caterpillar moth egg case, seven other moth caterpillars (of six species), and one sawfly larva.

stratagems. One of the most common is to lace their leaves with chemicals technically called allelochemicals or "secondary" metabolites (because they have no "primary" function in the plant's metabolism). These substances include poisons (cyanide in cherry leaves), drugs (nicotine, cocaine, caffeine), insecticides (pyrethrum, rotenone, and insect hormone mimics), and other substances that stunt or disrupt growth, kill, cause hallucinations or paralysis. However, no defense is perfect because each one has costs to generate and maintain. Additionally, each new defense eventually generates a new offense. If all tree species were defended with just one kind of toxin, then herbivores would probably easily evolve to tolerate that specific chemical. To reduce vulnerability, varieties of chemicals and tactics evolve as different trees specialize in their defense. In turn, insects specialize in their offense and select specific tree species. The more unique a tree's defense is the less likely an herbivore would be around to breach that specific defense. Using ants to kill or drive off insect leaf eaters, rather than having chemicals as a feeding deterrent, would appear to qualify at least early on in evolution, as a unique strategy.

I was irritated by having two days' construction work spoiled by ants. Now, with the passage of time, I can be more objective and concentrate instead on the fact that our attacker ants are interesting. Why did those ants insist on climbing that tree and attacking us? Where did they come from?

I suspect they came from a partially rotten pine stump about a hundred feet away. Some ten years earlier, I had cleared away the scrubby pine trees to create the clearing. Soon after the stumps were softened by decay, ants excavated them and made them into ant mounds.

These ants have become entertainment for me. On several occasions I have seen them engaged in their wars. They fight with a difficult-to-identify (or as yet unnamed) species that my friend the ant specialist E. O. Wilson kindly narrowed down to "one of the *Formica fusca* group." In the ant wars (described in my book *In a Patch of Fireweed*) the *F. subintegra* took the others as slaves. Once recently I also found the red raiders, *F. subintegra*, in mortal combat on the steps of the cabin with a colony of carpenter ants, *Camponotus* sp. Since the carpenter ants had previously invaded the lower logs of the cabin structure, I took more than a casual interest in the three-day battle. Each day the ground was littered with newly dead and dying carpenter ants. Sometimes the *F. subintegra* were still grappling with their feebly struggling victims. Maybe my son was right. The *F. subintegra* are "bad news."

F. subintegra are not only warriors, however. They are also nurturers. They tend aphids as carefully as the East African Masai warrior tribe herds its cattle. The ants keep the aphids only to "milk" their sugary secretions, as the Masai keep cattle for their milk (and some blood).

The insect herders with their aphids were common on the young shoots of quaking aspen. But I found them also on birch, alders, balsam firs, pines, and red spruce. While herding, the ants seemed to be perched almost motionless near the aphids, not running about at their usual frantic pace. When I looked closer, however, I found that they were gently palpating the aphids with their antennae. This palpation of the aphid "cow" is the signal that tells it to release a gob of sugary secretion, from its anus. The ant quickly takes the offered secretion but otherwise does not molest the aphid. Ants don't "kill the goose that lays the golden egg"—they keep coming back for more. Like the Masai warriors who unhesitatingly spear lions that threaten their cattle, herder ants execute insects alighting near them. They eat them for a protein supplement to their diet. (However, some caterpillars have evolved to take advantage of the ants' weakness for sweets: like aphids, they also secrete sugar, from specialized glands on the back and thereby buy protection from being eaten.)

Aphids are plant "parasites"—they feed on plants just as lice feed on mice, by sucking their fluids. Because aphids are tiny, they only cause problems when large numbers infest a tree. Therefore, it is preferable for a plant to tolerate the aphids and the ants, to have fluid taken from a twig rather than be overwhelmed be hordes of caterpillars and have all its leaves eaten. The "ant guard" that excludes leaf eaters is not free because it may admit aphids.

The price is nevertheless sometimes worthwhile. Bastiaan J. D. Meeuse, a distinguished botanist at the University of Washington, speculates that ants may have played a major role throughout evolutionary history in protecting trees from their insect predators. He cites the cocoa tree, *Theobroma cacao,* as a particularly instructive example of how ants help trees. In parts of Java and West Africa, a sap-sucking bug, *Helopeltis sumatranus* (in the same family as the blood-sucking bedbug), had made cocoa production nearly impossible by weakening the trees to a much greater extent than any aphids. To combat the bug the Dutch government in Java deliberately infected healthy young cocoa trees with aphids because the aphids were then immediately tended by the black Javanese ants *Dolichoderus bituberculatus*. These ants never tend plant bugs, such as *H. sumatranus*, because plant bugs don't exude sugar. But while tending aphids that *did* give sugar, the ants then eliminated the destructive bugs. Until then, these bugs had been compared with Attila the Hun.

Most of the small trees I found with aphid colonies were tended by at least thirty ants. It was impossible for me to keep track of that many individuals all at once. However, I found a more manageable group of only four or five ants on a two-foot-high quaking aspen sapling along the path to the cabin, and on an afternoon in late July 1996, I sat down beside it to do some serious ant-watching. One of the nice things about watching insects is that, unlike birds or mammals, insects seem to be oblivious to observers. They act as if they are in another world. There is little need to worry that their behavior is affected by your presence (unless they are of the kind that suck blood).

At first, I saw the five ants just running around, alternately palpating aphids with their antennae and then stopping to meet face-to-face with each other. When two ants met they palpated each other's heads as they had done the aphids' rear ends. Important transactions were occurring during these tête-à-têtes. Colony identities were being checked. Food was being transferred. Perhaps ants aren't perfect at discriminations. Maybe to an ant there is little difference between another ant's face and an aphid's rear (even though they surely recognize strange ants). Maybe the aphid's secret of survival with ants is that its rear end mimics an ant's front end because ants routinely share food to inform each other about colony nutrition. The aphids have found a simple solution for ingratiating themselves with a potential powerful enemy and making it work to defend themselves if need be. And they do it cheaply, using their own waste.

Within a half hour, two of the five ants with swollen gaster (abdomen) were running down the aspen head first, leaving in a hurry, ferrying honeydew, the aphid's sweet anal secretion, back to their colony. Two new ants had already run up the tree. A small green snout beetle was blundering along the ants' "highway," the thin tree stem. When the ant met the beetle, the beetle tucked in its legs and instantly fell to the ground like a small green stone. I then noticed a leafhopper on one of the poplar's leaves, and it leaped off into space when an ant came near. To my great surprise, I eventually detected a half-inch-long green inchworm (a geometrid caterpillar). It was motionless and aligned neatly along the lower midribs of a leaf, looking like the midrib itself. The ants did not crawl on the leaf undersides. Attached to the underside of another leaf was a frothy white gob of sticky spittlelike material.

Within the "spittle," I knew, would be a "spittlebug," or "froghopper." Adult froghoppers have wings and can jump marvelously far. Like the aphids, they also suck plant juices. Instead of giving up its surplus of plant juice to feed ants, as aphids do, a froghopper larva exudes and then surrounds itself with a protective froth that repels ants and within which it hides. This froth is not spit at all but an anal secretion. The larva turns this liquid into a froth by blowing exhaled air into it and adding a substance like soap that helps make a lasting lather.

After what seemed like an eventful half-hour adventure at the young aspen tree, I was encouraged to do a repeat observation the following morning. This time I'd keep an even closer watch on *individual* ants. It occurred to me that ants had come up the tree, tanked up on aphid juice, and left, but maybe that was just an illusion. Maybe what really had happened was that the same five ants had stayed there all the time working the aphids, while special "tanker" ants traveled back and forth between the nest after the tappers gave up their honeydew to their colony-mates.

During my second watch at the aspen sapling there were again five ants in attendance. Four of them were slow-moving and deliberate, patiently "milking" the aphids most of the time, but one individual acted noticeably different. This one, whom I shall call "the runner," was continually jogging all over the aspen sapling, occasionally stopping briefly as if marking time. This ant seemed almost oblivious to the aphids. It never stopped once to palpate them. After about ten minutes the runner engaged in a long tête-à-tête with one of the four resident aphid-milkers. Looking close I saw that it was indeed receiving a load of collected honeydew. Five minutes later the same thing happened. Immediately after this tank-up (I could see that it really was a tank-up because the ant's gaster was by now visibly swollen), it ran headfirst down the poplar shoot and hit the ant trail, presumably back to its nest. I was elated that my hunch about the tanker ants was right.

Ten minutes after the tanker had left, one of the other four ants was also becoming swollen with clear liquid honeydew. I could see its dark hard abdominal segments pulled apart to reveal the clear abdominal contents through its transparent membranes. I could look almost through the ant. No "tanker" ant had come in the meantime to relieve the rapidly ballooning ant, but this ant didn't wait. It, too, eventually ran down the poplar and headed back to the colony. The ants were not as inflexibly programmed as I had presumed.

A sawfly landed on one of the thirteen leaves of the poplar. When an ant eventually ran near it, it flew off. Two small green leafhoppers were perched immobile on the leaves' undersides, where the ants rarely walked. Several small black ants of another species stayed on the lowest leaves of the poplar and on the grass below it. Were they in hiding, perhaps lapping up the drippings?

Symbioses between ants and trees have been known for more than a century, and the intricate mutual accommodations between them indicate an ancient and mutually profitable relationship. The ant-tree associations are perhaps most important in the tropics and have there blossomed into obligate relationships where one needs the other to survive. Some tropical trees even provide ants apartments for living "on board." In acacia trees, for example, the large spines (that protect the twigs from mammalian browsers) are swollen and have roomy hollow bases that are the ants' secure nest sites. The ants then protect the leaves from *insect* predators.

Besides apartments in hollow spines the trees may provide additional enticement to ants. They also provide sugar bait from sugar-secreting glands on the bases of the leaves. These glands are, in effect, a substitute for aphids, who of course are universally known as plant parasites, except when . . . ants eat their sweet feces and they are then symbionts and partners instead.

MY WHITE PINES

*Strange that so few ever come to the woods to see how the pine
lives and grows and spires, lifting its evergreen arms to the
light—to see its perfect success.*
                    —Henry David Thoreau

The remnants of old hay fields just south of the ant-defended spruce
tree and aspens were rapidly becoming a solid stand purely of white
pines. Here, the red ants were far less successful in repelling intrusion. A logging crew came and thinned out the formerly dense
grove, providing space for maples, seedling oaks, beeches, and other
nut-bearing trees. A mixed forest is in the making, one that will
eventually be dominated by the great white pine, *Pinus strobus*, a tree
that was pivotal in Maine history.

Donald C. Peattie, in his classic book on American trees, maintained that "certainly no other [tree] has played so great a role in the
life and history of the American people." White pines once formed
huge stands over much of Pennsylvania, New York, and New
England. When the giant pines bloomed, wind swept the yellow
pollen over the sea where superstitious sailors presumed it was rain-
ing brimstone.

White pines were a dominant tree of the landscape when the Abenaki, the original Mainers, saw the first Europeans land on their shores. These huge pines or "goas" (pronounced "go-was"), as the Abenaki called them, were by far the biggest and oldest trees. Of all the trees in New England, the white pines (and possibly sugar maples) are the longest living, occasionally exceeding three hundred years in age. They may grow to a diameter more than 4 feet and a height of 150 feet. While their ages and dimensions are not impressive by standards of the western United States, where a bristlecone pine that is almost five thousand years old has been documented and where redwoods grow to more than 300 feet tall. In the Northern Forest, white pines stand out in many respects. The tree grows straight and tall. Its wood is smooth and soft-grained and possessed of a unique combination of light weight and strength. It is ideal for ship masts and lumber.

The first mill to make use of white pine was built in 1626 at York, Maine. From then on the "boom" of pine exploitation did not cease until virtually all of the old stands were depleted. Pines were shipped to England, Spain, South America and California, Portugal, Africa, the West Indies, and even to Madagascar.

Bangor, a seaport up the Penobscot River in Maine, became the greatest shipping port for pine lumber in the world, and its population soared from 277 people in 1830 (when Chicago was a hamlet of only twelve families) to 14,408 in 1860. Hundreds of sawmills were arrayed up and down the Penobscot, a river that served as the artery by which the trees were floated from the interior of northern Maine down to the coast. The Maine climate helped to expedite the boom and doom of its virgin pine stands. In winter the ice and snow and the solidly frozen ground made it easier for the oxen to drag the great trees to and onto the river, so that they could be floated to Bangor after the thaw in the spring. The pine boom spawned new breeds of men, the lumberjacks and the river drivers.

Maine paid homage to the white pine tree. Already in 1820 it was represented on the state seal, and in 1895 the state legislature designated its cone and tassel (the male flower) as the official Maine state flower. The tree itself was designated the official state tree of Maine in 1945, and so Maine came to be known as "The Pine Tree State." Despite the great pine boom that came to a bust when the original centuries-old stands were depleted, the white pine still remains common, especially in many regions of the central and southern parts of the state where it is now perhaps the predominant tree.

The felling of the pine giants on an unprecedented scale was alarming, and the dramatic "bust" following the pine boom activated the public. Conservationists who had fought for decades without allies suddenly had dedicated supporters. This arousal came too late to have much effect on saving the last virgin pine stands, but it did generate the will and the drive for Theodore Roosevelt and others to win conservation battles elsewhere. A new awareness of the finite nature of our forests helped bolster support for the National Parks and the Forest Service who was then more concerned with forest conservation than liquidation. The example of the pines' demise helped shape public policy not because it happened, but because it happened so suddenly. Far greater calamities, if they creep up slowly, are often hardly noticed.

Ultimately, however, it was the white pines' biology not public policy or restraint that was the most important factor in its preservation. In most cases, there is some small detail we are not aware of that we unwittingly violate in an ecosystem that then endangers it. The "little detail" of biology is that this particular species thrives on

White pine seedlings getting a start on open ground among grass.

disturbance. It was precisely the cutting down of the forest and the creation of pastures and burns and the scarification of soils that provided the ideal conditions for regrowth of this tree species.

White pine trees are shade-intolerant. Seedlings can only grow in a clearing. As a consequence, the young cannot grow under their parents. In the northeastern United States, clearings occur only where a disturbance has eliminated the trees. The clearing may be small, such as a hole created in the canopy of a recently fallen tree, or it may be large, such as that following a fire, logging, or the creation of a pasture.

Nobody knows how the huge aboriginal pine stands that covered vast tracts of North America were created. They must have been two hundred to three hundred years old, judging from the size of the trees they contained. Perhaps they grew in the paths of hurricanes, or in land left fallow by Indian farmers, or in the wake of great forest fires. Forest fires have been frequent in Maine. In 1795 a fire south of Mount Katahdin burned two hundred square miles. In 1825, the "worst" year on historical record (when I suspect my forest and the surrounding forests burned, see page 6), 5 percent of the state's forest acreage burned. From 1903 to 1954 there were 14,483 individual fires recorded. Fire favors pine regeneration under some circumstances because the large seed trees are much more resistant to fire than the competing softwoods that are killed.

Cleared ground is generally taken over by the survivors who are already there to deposit seed, so perhaps white pines get a jump start.

One thing of which we can be reasonably sure is that few seedlings would grow up where pine groves were already standing. The ancient virgin groves were likely temporary. They seemed permanent, but change was inevitable. The cycles and rates of changes would, however, be in terms of hundreds of years. The lumbermen who came to cut the trees, followed by the settlers who then cleared the land, inadvertently made new pine plantations when they left their land to go west. On my hill the pastures

A white pine tree about four years old.

were also left as the people who once lived here went west or moved to town to work in the mills.

Most of my pines grow in these abandoned pastures. They are now fifty to sixty years old. They still have smooth baby bark, but they already stand about fifty feet tall. The largest among them have a diameter at breast height (dbh in forestry lingo) of about twenty inches. Because of dense crowding as they grew, more and more of them died. Many were weakened and then died from an epidemic of white pine blister rust that raged through the stands. The white pine blister rust is a fungus (imported from Europe in seedling nursery stock) that leaves huge scabs on the trunks that profusely leak pitch before the trees die. The fungal spores invade through the leaves' (needles') air pores or stomates (through which the tree breathes). The epidemic subsided after I pulled up all of the wild skunk currants growing on the top of the hill. The fungus has a life cycle that needs not only pine trees, but also the agency of plants of the genus *Ribes*, to which skunk currants, gooseberries, and many other species of wild and cultivated plants belong.

The thinning out of the pasture pines by the fungus has not been all negative. The pines had annual growth rings of 0.22 inch per year, or an average annual increase in tree diameter of less than a half inch. The uncrowded trees now grow at rates two to three times greater.

The soil was undoubtedly poor after the sheep and cattle left the pine-studded pasture some fifty years ago. No one plant could dominate all the others on poor soil. There were probably a variety of grasses and meadow flowers. There were probably vesper and savannah sparrows, bobolinks, and perhaps eastern meadowlarks. There could have been a pair of kestrels nesting in an old hollow sugar maple, raising successive broods of young on grasshoppers and meadow voles. When I first came up this hill as a small boy in the

1950s there was already a quite different scene. There were many small bushy pine trees and blueberry vines, lycopods, and small feral apple trees. There were ground junipers under whose spreading branches the snowshoe hares hid. Deer and bear abounded, feeding on the apples in the fall. White-throated sparrows, hermit thrush, and Nashville warblers sang in the spring and built their nests here. All of this also came to an end. After a few more years the pines shot up and closed the canopy with their crowns, hogging all the direct light. One by one the apple trees died. The lycopods, the ground junipers, and the sprouting young hardwood trees also died. Soon the ground was barren. The hares left, as did most of the birds, the deer, and the bears. The forest floor was now a thick carpet of pine needles, and through these needles grew only the most shade-tolerant of plants, the *Maianthenum* lilies or Canada mayflower. In spring each plant poked up one broad green leaf in places where a little diffuse light still reached down. But there simply wasn't enough light for the one annual leaf to gather up enough energy for flowering.

In the winter of 1993–94, I made an agreement with the paper company in nearby Rumford (then Boise Cascade, formerly Oxford, and since sold to Mead Corporation) to have their forester oversee the cutting and marketing of trees from this forest. In return, the company got a small percentage of the profits and the first option to buy the wood for their mill. The local logger that Boise hired cut the trees we marked. We first cut those with blister rust, then those growing too closely together. No more than half the pines were taken and the ones that we left were usually the biggest. My first objective is not to harvest the trees, but to improve the stand by changing it from an artificial monoculture to a forest, a biologically diverse ecosystem.

Two springs after the selective cut I can already see a big change in the right direction. The slash (limbs and tops) has molded back into the ground. In the light gaps the *Maianthenum* lilies are growing in a thick green carpet that is speckled with white flowers in May. Some have red berries in the fall. The forest floor has sprouted a complement of hardwood seedlings, principally red and sugar maples, ash, black cherry, and quaking aspen. My seeding with acorns and other nuts should soon also result in a small scattering of oak, beech, and possibly walnut and hickory. Within a few years the white pine stand will become a mixed forest of several species and several vertical layers, and the remaining white pines will double and possibly triple their growth rate.

I am careful not to thin my pines out too much because young trees grow best in crowds where they support each other in storms. By removing too many, the remainder are endangered by windthrow. But a windthrown tree now and then is no disaster in this forest. The upturned roots with their clinging soil provide a foothold for young yellow birch trees. They provide nest sites for winter wrens, Acadia flycatchers, and (in wet woods) for the warbler called the northern waterthrush.

Young trees pay dearly for the mutual protection they get by growing in crowds. Often they exhaust their resources in the struggle to compete for light and so they perish as small undernourished beanpoles. *Helper* and *competitor* are relative terms. I leave "competitors" where some young trees have it too easy, to channel the energies of these trees toward reaching up and growing straight and tall. I leave "helpers" where the trees are not growing densely enough to shield each other from the wind.

Most trees have close to equal potential. Luck then determines the individual's fate. They are complex organisms that already know how to grow up. They need only the right environment.

Dense tangle of beanpole trees such as are created in a clear-cut or clearing.

Selective cutting in my pine grove has produced light gaps where a variety of trees are springing up, creating a forest from a previous monoculture.

In a forest white pines, surrounded by other trees, will grow straight and tall. They stretch themselves toward the light. Those that grow in unnatural isolation, such as in a pasture, reach out to all sides. Before other trees crowded in, many of my pasture pines did just that. They grew huge lateral limbs, and their crowns were often branched, as if, because of an abundance of light, they did not need to focus on growing tall. In the winter the far-spreading lateral

branches become overburdened with ice and snow, and many of these great limbs and some of the crowns broke off. These trees now grow crooked and misshapen. I want some of my pines to grow huge and tall, so in those areas of the old pasture where the trees are not trained by the competition of the forest, I exercise my pruning saw. By snipping a side branch here and there I can prevent the growth of a double trunk that would bend and break the whole tree. By snipping off a few lateral branches I force the tree to reach up. By removing closely crowded competitors, I give some trees a chance to grow as large as they can.

I measure my pines not against western American standards, but against the thickest, tallest pines I've seen in Franklin County. One of them grows in a stand of mature red oaks, white and yellow birches, and maples, between Perry Pond and Perry Mountain about twenty miles from my cabin. This pine, like most pines, grew from a previous forest disturbance, because, as mentioned already, pines are shade-intolerant and cannot grow under the canopy of a mature forest. The forest of large trees by Perry Mountain was a pasture once, too. This pine persists now within the matured hardwood forest because it grew straight from its youth. Now its crown reaches over the forest, capturing the sunlight.

The first time I walked past this tree on my way up to the cliffs of Perry Mountain to examine a raven's nest, I thought, "Hmm—a big tree," and let it go at that. The tree stayed on my mind, and months later I came back to take another look at it. This time I noticed that the lowest limbs were all dead. There were no live limbs up to about forty or fifty feet. The dead limbs were thick and reached ten to fifteen feet to the sides, off into the hardwood forest. The pine had therefore started to grow along *with* the hardwood forest: there must have been sunlight available out to the sides of the tree to permit such lateral growth. The forest had eventually overtaken the limbs, but the pine had kept on pace, reaching even higher for the sun, always keeping itself above the rest of the forest by adding new growth at the top.

Each pine branch is an economic probe into the world. As they grow laterally, some probes encounter space and light. They respond quickly, branching profusely, thickening, and then reaching still far-

A large white pine that started in mixed forest, then reached over it.

An old white pine that grew up in the open, without competition. It spread laterally in all directions and suffered much damage from ice storms.

ther out. Other limbs encounter shade. Apparently recognizing a losing proposition the tree reduces further investment in those limbs. They stop growing and die. It is this constant self-pruning that makes the pine tall, strong, and straight. Without self-pruning a pine could survive only if it lived in a world without competitors. It could not grow tall.

On my third visit to the big pine, I brought along friends. We admired the tree together and asked it a few questions. How tall are you? How thick? How fast are you growing? The tree was near the end of its life and it was no longer growing fast. With a small tree borer we took a 2-inch "core" from the trunk. The latest growth rings averaged only 0.11 inch. At that rate of growth the tree should increase its 40-inch diameter (we measured a 123-inch circumference) only another 22 inches in a hundred years, a growth rate close to that of my young crowded pines.

We were unable to climb the tree to measure its exact height, but by using a bit of geometry (at a specific distance from the tree we measured the angle to its top; height = the tangent of that angle × distance) we estimated that it was just barely 100 feet tall. It was a large pine tree according to an old (1896) book called *The White Pine*, by Gifford Pinchot and Henry S. Graves (Pinchot was twice governor of Pennsylvania and, from 1898 to 1910, chief of the U.S. Forest Service). In the book the authors explain how to measure a tree's increase in board feet over a set number of years. They single out a record white pine in Pennsylvania that was 351 years old and 155 feet tall, with diameter at breast height of 42 inches. This tree contained what they estimated to be 3,325 board feet of merchantable lumber. Records for southern New Hampshire in "The Virgin Upland Forest of Central New England" by A. C. Cline and S. H. Spor (*Harvard Forest Bulletin*, no. 21) list the record for white pine as a 48-inch tree, some 150 feet tall and 280 years old. The next largest species was hemlock, with the record tree at 36 inches girth and 110 feet tall. The pine at Perry Pond was thus close to record size by the standards of 100 years ago. I suspect it is a nearly mature pine, since two others near it had recently died of natural causes.

This pine would have pleased George III, king of England. More than a century before the book by Pinchot and Graves, New England did not belong to the Indians or to the settlers. English law stated that all oak and pine in the colonies 24 inches or more in diameter at the base, standing within three miles of a river, and not in private hands, were reserved for the Crown. English warships such as the classic 74-gun "ships of the line" each incorporated 2,600 tons of timber in the hull alone, which required about 700 large oak trees. The three masts of one of these ships were each 120 feet tall, absolutely straight, 40 inches wide at the base and at least 27 inches at the top. It staggers my imagination to think that the impressive pine by Perry Pond would have been too small to use as a mast. Nowhere in Europe were there trees with dimensions suitable for use as masts for the giant battleships. The king's agents marked a Broad Arrow, three blows of an ax, on the trunks of "mast trees," to indicate that they were to be reserved exclusively for the Royal Navy and to serve the British Empire.

There was trouble with the king's law. Because large areas of land were unsettled, there was dispute over how to define "private lands" and the king's right to these lands was questioned. Nevertheless, the king's agents paid good money for securing mast pines and men searched for them. After finding one, they would make roads to it through the wilderness. Twenty-ton trees were hoisted onto three pairs of wheels, then dragged by up to forty oxen in teams of two, for as long as twenty miles to the banks of the Androscoggin, Kennebec, and Saco Rivers of Maine. Specially designed ships then ferried the cargo to England.

In the fall of 1775, the British bombarded and burned Portland (then Falmouth), Maine, because they became incensed by Mainers' opposition to the delivery of pine masts exclusively for the king of England (they often sold them to England's enemies instead). In the same year fighting erupted at Lexington, Massachusetts, interrupting the harvest of trees, and the mast ships at Portland and Portsmouth sailed back empty.

For a long time after the Revolution, there were still pines marked with the Broad Arrows in the woods. These pines now supplied the American navy. Meanwhile, the boards in many old houses, including the farmhouse where I lived, in Maine were frequently twenty-two to twenty-three inches wide, but presumably never the incriminating twenty-four.

We now have jet airliners, and spaceships that fly to the moon with rockets. However, the much larger warships, with their twenty-ton tree masts, sailed the world's oceans powered only by the wind. Now, looking at this old pine I not only see the beauty of the live tree but also hear the echoes of bygone rousing seafaring chants and smell the salty spray. Even now, Maine's own "Schooner Fare" folk-singing group sings of sailing ships: "And all that matters, is the wind upon the Main [sail] and the masts will rise again." I wonder: Will they?

Will our fossil fuels soon run out (say, in a generation or two of pine trees) or will they still be available for many more centuries? Will fossil fuels become so expensive that we will turn back to the tall pines to hold up the sails to harness the winds? My pines will take one hundred to two hundred years to grow old. I'd like to think of them dying of natural causes. However, I cannot escape the thought of the large-scale devastation of ecosystems a large human population would inevitably cause when it exhausts a finite energy supply. Using the power from the sun as collected by trees, to harness the wind, is not a bad alternative. Raising pine trees for masts is not, however, written into the Forest Management Plan that I'm annually required by law to submit to the state office at Augusta.

A Celebration

In the woods, too, a man casts off his years, as the snake his
slough, and at what period soever of life, is always a child.
. . . In this wood, we return to reason and faith.
—Ralph Waldo Emerson

I'm at the door of my cabin at dawn, looking east over the forest toward Alder Stream. Frost lies white and heavy upon the grass. It is one of the first frosts of the year, coming on the first weekend in October. It is a welcome and expected announcement of the finality of one season, and the beginning of another. The sun has not yet come over the ridge, and I banish the morning chill with a cup of coffee—mesmerized by the peaceful dawn. As the sky's reds, purples, and yellows emerge and then fade to blue, the maples' yellows, oranges, and reds become vibrant out across the clearing at the edge of the forest. All is silent except for the morning calls of Goliath and Whitefeather, my tame raven pair that live free on the hill. Within a minute or two, I expect they'll fly out over the cabin to pursue their errands in the forest. Today is our day of the "annual gathering." Many of my friends will climb the hill to join in the festivities later in the morning. A few are already here, having camped last night in the field.

A crash in the pine grove nearby brings me out of my reverie. Within a few seconds the source of that crash, a bull moose, majestically emerges into the clearing. He stops briefly, then slowly walks up toward the cabin. I'm in plain sight by the door, coffee cup in hand. His dark eyes regard me only briefly and without apparent interest, as he walks slowly and nonchalantly to within ten feet of the apple tree where the cow and her calf had previously nibbled (page 129). Then he ambles past a guest's tent and stops next to the maple syrup evaporating pan that is already set up to boil lobsters. He continues on to the fire pit where we'll broil sixty pounds of steak this afternoon. I yell to Bruce and Rosemarie's tent: "Moose coming!" The moose ambles past the tent and back into my thinned-out pine stand. I follow for a while, strangely moved by this uninvited but very welcome visitor at our gathering.

It has now been twenty years since I first bought this land. When I first came here, large portions of it had recently been cut over and other parts had been cow pasture. That is why I paid such a low price. The old fields were overgrown with scrubby young pine. The orchards were a tangled brush of red and sugar maples.

The same old fields, into which this moose now wanders, have already yielded their first pine logs. Once again the light reaches the ground, and a mixed stand of maples and other hardwoods is springing up. In the light gaps where trees have been removed there are now acorns. Small oaks will begin to grow next spring. Large pines will stand here in a mixed forest in ten to thirteen years. Wood warblers and hermit thrush have returned. The remaining pines, in turn, are less crowded and are now taking off in a renewed rate of growth.

Other portions of the property had been dominated by stands of mature balsam firs, by quaking aspen or red spruce. Many of these have been thinned out. A selective harvest like the last will again be possible in fifteen to twenty years. There was no clear-cutting, no herbiciding, and there will be no planting of "tree plantations" of the sort that the forest industry often refers to as necessary for "sustainable forestry."

The economic benefit of my enterprise has been significant, if not astonishing to me. My cash profit has already twice exceeded the original price of the property plus all expenses (the taxes I've paid in twenty years). This amount of profit was all "stumpage" (lumber sold "on the stump"). The fir and aspen that were cut are now paper. The spruce and pine logs are boards. The birch logs are veneer. All these products are being passed on to countless other industries and human uses. The money that I got is only a very small percentage of the total dollar value of the timber, which largely benefited others. Most immediately it provided work for nine local loggers. The harvested trees, in turn, provided work in local mills. Every day I feel the contentment of having the opportunity to "live off the land" and getting to know it and become connected to it. Every year I see the trees grow into a living forest.

In thanks for all of these bounties I have given an acre of the land with the original deer camp (which the original owners called Kamp Kaflunk) back to an Adams, of the family on whose farm I first did chores, chased raccoons at night, and threw apples with a stick into the farm pond. I gave the camp to Bill as a birthday present, and I regard it as a token of my appreciation to the Adamses for what they gave me at their farm forty-seven years ago. Forty-seven years have passed since Bill and I climbed the apple tree next to the shed with the bantam chickens and ate the early yellow goldens or flung them across the stone wall.

This weekend we celebrate the past and our hopes for the future. Friends have come to renew old ties and establish new ones. Bill

Adams is here, and more than a hundred other people came from eight states. One came all the way from Madras, Oregon, on his bicycle. Indeed, Gary was the first to spot the moose this morning.

The moose has wandered off and two hours later the main body of the guests arrive and gather by the cabin, especially around the beer keg. "Let's go cut some maple switches," I suggest. Bill picks up the cue: "Cutter," he says to his son, "go pick a bushel of them apples," motioning to the old tree next to the cabin that had attracted the moose.

We're about to begin the annual apple toss. Having secured the accoutrements, detailed instructions follow: "Be sure you have a stick that is flexible but firm. . . . Put the apple on tight so it will hang onto the stick as long as possible when you give it that hard toss. The harder it hangs on, the more force you can put behind it, and the more power will be packed into it when it does let go and fly."

"Hey, Bill, you take the first shot. Show them how it's done. . . ." And as an afterthought I add, "Just make believe that the frog pond is down there at the end of the field. . . ."

Bill smiles, picks up an apple, impales it onto his stick. "Here goes!"

Slowly he swings it over his shoulder, then whips it forward over his head in a smoothly accelerating motion, then jerks back in a quick flick just as the apple is at maximum velocity. The force of his muscles, magnified by the leverage of the stick, all bear down on the apple as the switch is stopped short. The apple is released into the air. It flies at first in a long straight line, gaining altitude on its way across the field. Slowed by air resistance and tugged by gravity, it settles into a long graceful arc and lands in the red maples on the other side of the clearing. We hear the missile pounding its way through the foliage before hitting the forest floor.

"Ya*hoo*!" he yells, and everyone cheers.

The contest is on. Who knows. Perhaps a seed from one of the apples we tossed today will germinate and grow, providing fruit for the grandchildren of that moose who visited this morning and for my newborn son when he is a teenager.

Appendix A

THE TREES FOR THE FOREST

Thou canst not stir a flower
Without troubling of a star
——Francis Thompson

When I bought my three hundred acres ("more or less") twenty-two years ago I had no thought of "managing" a forest. I wanted a retreat, a cabin site. The rest was a tax burden and therefore something to unload quickly. But I got to know the land, and I became attached to it. Parts that had once been pasture started to become forest. I decided not to sell. It is still all registered in my name, and because of a recent Maine law, I'm required to submit a "forest management plan" to the state office in Augusta. I could not totally sidestep the issue of forest management, and it seemed appropriate not to do so in this book either, because my forest is part of a much larger forest the fate of which is one of the great issues of our times. Most of life on this planet depends on forests and worldwide they are receding at an alarming rate, being replaced with sterile tree plantations, cornfields, cow pastures, shopping malls, cities. . . .

Forest management is a complex issue with many perspectives. Management options depend on the specific ecosystem involved, on long- vs. short-term views of what the ecosystem should contain, on previous land use, on scale, on values, on species composition, on logging equipment, and on long- vs. short-term economic incentives. Looking from one perspective only is a sure recipe for mayhem. I shall try to bracket the different considerations somewhat by focusing on what works for me in my forest, and why I think it works in terms of my values and objectives.

What I had read and heard about "forest management" had often only confused me. I needed to see with my own eyes. Some of my land had been pasture and some clearing. Now there was thick regrowth. I started by thinning this regrowth. My motivation was not to earn money, but merely to remove single-aged, short-lived trees (aspen and fir) that were mature and falling in droves. In other areas young white pines predominated. These long-lived pioneering trees would dominate the site for centuries, and since one of my goals was species diversity I selectively harvested them to give other species a chance to grow up in the understory. In general I was very pleased with the resulting "look" of the land. It was returning to a diverse forest. On my land selective harvesting encouraged species diversity. I therefore experimented further.

My next "experiment" was with method. I sold stumpage (wood "on the stump") to three different types of loggers. The first was a horse logger who claimed he would do a neat clean job. The second was a man with a skidder, a machine for dragging trees. And the third was a mechanized crew using the combination of a feller-buncher, grappler-skidder, and delimber. I had expected the most damage with the latter, large-scale operation. However, I was wrong in my assessment. The feller-buncher with its long boom could reach far into the woods in many directions from a narrow path. It could individually grasp, saw off, then lift and pile trees without moving from the spot. A skidder may trample all around destroying many young trees and scarring others in the process of harvesting the selected mature trees. The grappler-skidder "grapples" whole piles of logs and "skids" them on a long narrow trail through and out of the woods to a "landing" where they are hauled out. The horse logger could drag only one or two stems at a time, and then not very far. He needed to bulldoze a road in to where a truck could haul the logs out! I was surprised that the most equipment-oriented method was not *necessarily* the most destructive, at least at this site.

The obvious drawback of the mechanized method may be in the long term. The small operator can live well by cutting less wood or by working longer on a given woodlot. Hay for the horse is not a big overhead cost. In contrast, the big operator finished my job in a couple of weeks, then he immediately had to go on to the next operation in order to be able to make the huge monthly loan payments on his multimillion-dollar equipment. He needed to cut huge amounts of timber quickly just to eke out a profit above and beyond paying for his equipment. I suspect also that in the long run the mechanized operation will take jobs away from local people. It does not encourage local people with local means to have local jobs. One huge machine replaces many loggers—people who have a stake in a future, while the machine has none.

The method by which the trees are cut is only one issue. Choosing trees to cut is another. I've specifically left some of the largest, still fast-growing trees, taking instead those that were damaged, dying, or densely crowded. Possibly the same formula would not apply to all forests, but selective cutting works for me. It works ecologically because the forest (which was not old growth to begin with) is now much more diverse. It works economically because I've more than doubly paid for the cost of the land plus taxes. It was all "pure" profit to me. It provided work for innumerable people in addition to loggers, and it provided the wood we all need.

I now allow the forest to grow what it "wants" to grow. Cedars do well in the low area by the brook. Red spruces thrive on the rocky ridge. Sugar maples, white ash trees, and white pines abound in and around the old pastures. Diversity and competition ensure that each spot will grow what is best there, and in the long run diversity is economically sound. It is a buffer against fluctuating markets and potentially devastating tree diseases and pests.

In providing a forest "management" plan to the state (in order to get a tax break for using the land to grow wood) it seems to me that one plan is to do nothing—just leave it alone. This type of land use, rather than intensive harvesting, is perhaps also the most socially and ethically desirable and therefore ought to be encouraged through tax breaks. Management, of which the *most* intense is tree planting on tree plantations, should be discouraged rather than encouraged by taxing. The very idea of "managing" a forest in the first place seems oxymoronic, because a forest is an ecosystem that is by definition self-managing. Calling the growing of wood on plantations "forest management" is the same as defining the farming of corn in Iowa as "prairie management."

Here in Maine I see the start of a revolution in our use of the forest land, much as there once was an "agricultural revolution" during which vast areas of the earth's most productive lands were gobbled up for growing crops. Where will it stop? In America's midsection, a few settlers tilled the soil and no harm was done, at first. Eventually, a prairie ecosystem spanning thousands of miles was put to the plow. Soon nothing was left but a pampered corn plant, endlessly cloned (where before there had been hardy native grasses, richness of species, and vast herds of buffalo). Much of the corn is used to feed cattle in distant rangeland where these domestic animals are degrading still other native habitat. Given the choice, I'd rather eat bison shot on the wide open prairie, and raised on buffalo grass among which the savannah sparrows nest, the meadowlarks sing, and the upland plovers cavort. Similarly, I'd rather use wood from forests than from a tree plantation.

Plantations are generally seen as a step above clear-cutting, but the two are linked. In practice, it is not always clear which one is driving the other. Clear-cutting yields the maximum profits in the *immediate* short-term. The rank growth of commercially worthless raspberries, pin cherries, and striped maple after a clear-cut can still be tolerated as long as aerial herbicide spraying is an option; it permits replacing the clear-cut with plantations that yield more profit in the *near* short-term.

The other economically viable short-term option for a commercial enterprise such as a paper mill is to sell the land (the paper mill closest to my forest, in Rumford, has changed ownership at least three times in the last twenty years), because the mill's forest land can, under current law, be considered a business expense and tax write-off for six years. After the six-year tax write-off is no longer applicable, there is incentive to clear-cut and sell it to the next sucker as *his* tax write-off. Given this "state of nature," as in the fisheries industry, mutual destruction of all major competing resource users is assured in the long run if each pursues his best interests of the moment, unless limits are imposed to maintain sufficient biological *capital* needed to produce harvestable "interest" for the long-term survival of all.

Since the dawn of agriculture farmers first claimed the most fertile land, leaving the rest as forest. Now industrial foresters, after extensive clear cutters, want the rest (well, a little at a time), for tree plantations. So far there are not many tree plantations in Maine. But the scary part is that when you go on paper company tours they *show them off*, rather than hide them. Why? Because tree planting is touted as good for "sustainable forestry." Some people even claim it's more "efficient" to grow wood that way, while ignoring all the stronger arguments against it. As for me, given what I've seen in my forest, I'm not even convinced of the economic argument. But if I were convinced, the reasons *against* are compelling in their own right.

A plantation consists of even-aged trees of one species growing in rows. There is only one dense layer of leaves to catch the sun's energy. Below that layer nothing grows. It is a dead zone. In contrast, in a *forest* there are many layers, including shade-tolerant species that grow below that canopy. Many species can grow in a forest, not just the one chosen for present economic use. (Currently fir and spruce are much favored, but timber prices are highly unpredictable. Prices for a given species are often low one year and skyrocket the next.) In a forest, where different species thrive because each has a slightly different optimum for light, acidity, drainage, temperature, and minerals, each spot of ground gets exploited by the one best suited for that niche, and through diversity, one's "investment portfolio" is more secure in case a species chosen now no longer brings top dollar fifty years from now.

Growth of the trees—of *all* together—is better as well. The forest is an ecosystem. And in an ecosystem there is mutual interdependence, where the waste that one produces, or the resources that one doesn't use, becomes food for the other. A forest with high biodiversity is much more resistant to attacks by insects and other pathogens than is a plantation that is a sitting "target" for the right pathogen to come and wipe it out. Given all of these considerations, how can an ecosystem *not* produce more wood and potentially more valuable wood in the *long term*—than a uniform monolayer imposed on a nonuniform environment? (The *individual* plantation trees often have much thicker growth rings than forest trees, because in a forest the growth is spread among *other* large trees as well. It is these other trees that are tall and ready to bulk out quickly when other trees are removed. There is no need to begin again at the seedling stage, where annual wood accumulation is minuscule.

Unlike a tree plantation, my forest contains an integrated, highly interdependent mix of species that grows best at all the different climatic conditions and aspects of drainage and soil that are unique to my hill. The hill selects them. This forest contains tens of thousands of species of insects, thousands of fungi, dozens of birds. It contains viruses, bacteria, nematodes, tardigrades, and even a few amphibians and reptiles. Swallowtail butterfly larvae feed on its cherry leaves and the adults pollinate honeysuckle. There are water shrews in the mossy undercuts of the rivulets running into Alder Stream. There are wood frogs, yellow-spotted newts, yellow-bellied sapsuckers, and brown creepers. This forest is but a tiny part of the much more extensive great Northern Forest. Together this forest is a loosely amalgated "superorganism" whose interdependent parts function as a seamless whole. It channels or can channel materials from the soil and the atmosphere and energy from the sun, into and through a vast assemblage of life from solitary vireo to salmon.

A tree plantation grows wood as a crop. Period. Crops such as beans, wheat, and corn are genetically altered plants that can no longer exist in ecosystems. We have destroyed a major portion of the earth's ecosystems—prairies, wetlands, forests—to grow crops. So far trees have not been genetically altered much to make them pure crops. I shudder at the thought of more progress in this direction, because in the long term it would mean they could no longer survive in a natural forest. Progress in this direction ultimately, and perhaps inevitably, spells obliteration of even more forest to make more plantations.

Large-scale plantations can *only* be made after first clear-cutting. Helicopters are then often used to strafe the land with nozzle guns spouting "herbicides." These so-called herbicides kill raspberries that *only* grow in clear-cuts. Principally, however, they are designed to kill *trees*—young regenerating deciduous trees. Herbiciding for "forestry" is a deforestation technique adapted for creating lebensraum for a given species. Often it is a genetically uniform, non-native conifer planted in rows. (Paper can be, and is, made from both conifers and hardwood trees. Consumers could choose but are generally not given a choice.) There has already been public outcry over the clear-cutting and the spraying, but as I'll show, tree plantations are much more insidious and ultimately more damaging than either.

Spraying to kill trees and raspberry bushes after a clear-cut merely looks unaesthetic for a short time, but tree plantations are deliberate ecodeath. Yet, tree planting is often pictorially advertised on television and in national magazines by focusing on cupped caring hands around a seedling. But forests do not need this godlike interference. The biggest problem is not clear-cuts per se, but what's done with the land afterward. Planting tree plantations is *permanent* deforestation. Monocultures of western Sitka spruce do not belong in England, or Australian eucalyptus in South Africa, or red or Scots pine in Maine. All replace ecosystems, sometimes on a vast scale. The extensive planting of just one exotic species removes thousands of native species.

In a monoculture the trees *must* be managed like an agricultural crop. Competitors and predators are killed by herbicides and pesticides. After they reach maturity it makes sense to harvest them all at once, as one harvests a field of corn. Thus, the cycle is repeated for yet another series of decades. Meanwhile, the land is ever-more irrevocably sterilized. The end result is ecosystem destruction that may be well-nigh irreversible.

As one example of a global trend that seems to confuse the trees with the forest, I turn to one supposedly forested country, Finland. Finland, recognizing the importance of a continual wood supply, masquerades a gross vice as a virtue by actually advertising itself as a model country for "sustainable forestry." Like Maine, Finland is officially classified as about 87 percent "forest." As in Maine, the forest in Finland had in the past always been the economic and cultural backbone that supported and sustained the people for untold generations. As in Maine, pulp and paper industries moved in and took over. Now, for every (native) tree that is felled, two (exotic) trees are *planted*! As a direct result of "tree planting" Finland now has virtually *no* forest; 98 percent of its tree growth is now in even-aged monocultures of exotic Scots pine and Norway spruce. As a direct consequence of this deforestation, half of Finland's species of native forest plants and animals are endangered, and rapid inroads are being made even into the few remaining pockets of old growth wilderness where native species remain.

Why do we need forest? There are many compelling reasons, but they are not generally tabulated in quarterly balance sheets. Ecosystem "services" include air and water purification, flood control, erosion control, nutrient recycling, and pollination services. Ecologically, forest is natural habitat and most species on earth are adapted to it and require it. There have been and are many practical arguments for forest (wilderness)—practical in the material sense. Ultimately advocates for forests are motivated also by ethics rooted in our own psychological dependence on wilderness. Conservationists such as Aldo Leopold ("a thing is right—"), Connie Barlow ("because it is my religion"), E. O.

Wilson (biophilia hypothesis), and John Muir; authors such as Thoreau ("in wilderness is the preservation of the world"); and politicians such as Teddy Roosevelt, who set up a new concept on this earth (national parks), have all in their own way reaffirmed that like other animals we, too, need forest or other wilderness to live healthy adapted lives. Like other species, we evolved in wilderness and although we are now able to satisfy many of our *physical* needs outside it (at least in the short term), psychologically we still need the vital diversity, complexity, grandeur, and beauty of wild places. We need to feel *connected* to something tangible that can be seen, smelled, tasted, that is much greater than our own fleeting existence. Call it religion. There are untold millions who believe in this religion although they may not come out and say so, mostly because its tenets are so deeply ingrained as to be taken for granted. No *organized* faith is therefore necessary. To me and others of this religion a wilderness made by God and/or by the mechanism of evolution is at least as, if not more, holy than a cathedral made by man, and to harm it is a desecration. I see enough glimpses of this wilderness in my forest to feel inspired by a feeling of interconnectedness with the web of life. It gives me a dream. It is a realistic dream that is not destructive, and that all can take part in and enjoy the results. Preserving and fostering the fantastic life on earth grants infinitely more practical and intellectual rewards than the expensive but trivial knowledge of whether there are microbes on Mars.

How can we maintain our forests? Perhaps there is a lesson in the example of the Sundarbans, the great mangrove swamp wildlife reserve that stretches between India and Bangladesh along the Bay of Bengal. The Sundarban is the largest tract of mangrove forest in the world. It also has the largest population of tigers in the world. The two facts are interrelated, because the tigers are man-eaters and they exert a high tax on forest use. They exert "ecodiscipline," as Sy

Montgomery writes in her wonderful book *Spell of the Tiger*. Without its tigers the forest would long ago have been destroyed by people. The forest is retained only because the tigers, by hunting people who trespass on the reserve, effectively reduce the crush of human expansion. The tigers, not the government, are the keepers of this forest that nourishes many people.

To use the Sundarban forest in India you must pay a tax—the perhaps one-in-a-thousand chance that you'll be eaten by a tiger. A suburban person in Boston using the forest (say by building a house, using a shopping bag, buying a book or a newspaper) is too far removed from the cost of maintaining a forest to pay directly, although he or she may be more than willing to pay that cost. That cost should not be a punishment or a threat. Neither should it be borne only by generous or enlightened people willing to pay. It must be borne equally by all, and in direct proportion to their individual use of the forest.

We obviously cannot put tigers into our forests so that woodcutters would demand high prices, which they would then pass on to us, the ultimate forest users. Tigers won't do here. But protecting a "spotted owl" or some other creature, that, like the tiger, is also diagnostic of ecosystem health, might. It would guide harvesting. Supply-and-demand economics would then justly apply to all users.

In nature, the evolution of spectacular creations has resulted not from laws or decrees, but from the incessant application of consistent selective pressure acting on individuals. Similarly, in society we can create our own "selective pressure" through economic incentives to promote and maintain our biological heritage and to encourage its survival. There is a famous Jewish proverb that if you save one life, you save mankind. Similarly, by saving one species of animal or one piece of forest, we save the world.

Appendix B

THE TREES AND SHRUBS OF MY FOREST

Coniferous Trees

Balsam fir *Abies balsamifera*

Red spruce *Picea rubens*

White pine *Pinus Strobus*

Hemlock *Tsuga canadensis*

Larch/tamarack *Larix laricina*

White cedar/arborvitae *Tsuja occidentalis*

White spruce *Picea glauca*

Black spruce *P. mariana*

Large Deciduous Trees

Red/soft maple *Acer rubrum*

Sugar/rock maple *A. saccharum*

Quaking aspen *Populus tremuloides*

White/paper birch *Betula papyrifera*

White ash *Fraxinus americana*

American beech *Fagus grandifolia*

Yellow birch *Betula lutea*

Black cherry *Prunus serotina*

Big tooth aspen *Populus grandidentata*

Northern red oak *Quercus rubra*

Small to Medium Deciduous Trees-shrubs

Speckled alder *Alnus rugosa*

Striped maple/moosewood *Acer pensylvanicum*

Gray birch *Betula populifolia*

Pin/fire cherry *Prunus pensylvanica*

Chokecherry *P. virginiana*

Mountain ash *Sorbus americana*

Mountain maple *Acer spicatum*

Juneberry/serviceberry/shadbush *Amelanchier* spp.

Eastern hop hornbeam/ironwood *Ostrya virginiana*

Rare Trees

Black ash *Fraxinus nigra*

Butternut *Juglans cinerea*

American basswood/linden *Tilia americana*

Silver maple *Acer saccharinum*

Balsam poplar *Populus balsamifera*

American elm *Ulmus americana*

Coniferous Shrubs (Rare)

American yew *Taxus canadensis*

Juniper *Juniperus virginiana*

Deciduous Shrubs

Lowbush blueberry *Vaccinium* spp.

Beaked hazelnut *Corylus cornuta*

Witch hazel *Hamamelis virginiana*

American elderberry *Sambucus canadensis*

Red elderberry *S. racemosa* (*pubens*)

Skunk currant *Ribes glandulosum*

Red osier dogwood *Cornus sericea*

White-fruited dogwood *C. foemina*

Alternate leaf dogwood *C. alternifolia*

Rose *Rosa* spp.

Meadowsweet *Spiraea latifolia*

Steeplebush *S. tomentosa*

Hobblebush *Viburnum alnifolium*

Arrowwood *V. dentatum*

Nannyberry *V. lentago*

Maple-leaved arrowwood *V. acerifolium*

Highbush cranberry *V. Opulus*

Wild grape *Vitis* spp.

Honeysuckle *Lonicera* spp.

Willow *Salix* spp.

Planted/Introduced Trees-shrubs by Cabin

Black locust *Robinia pseudoacacia* (two)

American chestnut *Castanea dentata* (four)

Pear *Pyrus communis*

Apple *Malus sylvestris*

Lilac *Syringa vulgaris* (two)

Hawthorne *Crataegus* sp.

Horse chestnut *Aesculus* (one)

Black walnut *Juglans nigra* (seeds)

Bitternut hickory *Carya cordiformis* (seeds)

Shagbark hickory *C. laciniosa* (seeds)

REFERENCES

The following references provide additional information.

Allen, Michael, F. 1991. *The Ecology of Mycorrhizae*. Cambridge Univ. Press., Cambridge.

Alverson, William S., Walter Kuhlmann, and Donald M. Waller. 1994. *Wild Forests: Conservation Biology and Public Policy*. Island Press, Washington, D.C.

Attenborough, David. 1955. *The Private Lives of Plants*. Princeton Univ. Press, Princeton, N.J. Hardcover with extensive color. This book skillfully picks the highlights of plant adaptation.

Bechmann, Roland. 1990. *Trees and Man: The Forest in the Middle Ages*. Paragon House, New York.

Beek, Donald. 1977. Twelve-year acorn yield in southern Appalachian oaks. *USDA Forest Service Research Note SE–244*.

Berry, Wendall. 1995. Private property and the common wealth. *Wild Earth* (Fall): 4–12.

Botkin, Daniel B. 1990. *Discordant Harmonies: A New Ecology for the Twenty-first Century*. Oxford Univ. Press, New York and Oxford. Botkin stresses the dynamism of nature, questioning the concept of "stability" for climax ecosystems.

Colborn, Theo, D. Dumanoski, and J. P. Myers. 1996. *Our Stolen Future*. Dutton, New York. Some critics say this is a sequel to Rachel Carson's *Silent Spring*. These authors explore how man-made chemicals from some industrial processes affect behavior and development at extremely low doses, long before they poison adults.

Connor, Sheila. 1994. *New England Natives: A Celebration of People and Trees*. Harvard Univ. Press, Cambridge, Mass. This book demonstrates the importance of wood in New England history. Along the way it gives detailed accounts of many species. With color plates and many photographs. A good reference source slanted to New England culture rather than forests.

Coolidge, Philip T. 1963. *History of the Maine Woods*. Furbush-Roberts Printing Company, Inc., Bangor, Maine.

Coutts, M. P., and J. Grace, eds. 1995. *Wind and Trees*. Cambridge Univ. Press, Cambridge.

Daily, Gretchen C. 1993. Heartwood decay and vertical distribution of red-naped sapsucker nest cavities. *Wilson Bulletin.* 105: 674–79.

Daily, Gretchen C., Paul R. Ehrlich, and Nick M. Haddad. 1993. Double keystone bird in a keystone species complex. *Proceedings National Academy of Sciences USA* 90: 587–94.

De Vries, Philip J. 1992. Singing caterpillars, ants and symbiosis. *Scientific American* (October): 76–82.

Dobbs, David, and Richard Ober. *The Northern Forest.* Chelsea Green Publ. Co., White River Junction, Vermont. Mostly about the citizens of the northern forests.

Donaghue, M. J. 1994. Progress and prospects in reconstructing plant phylogeny. *Annals Missouri Botanical Garden* 81: 405–18.

Ehrlich, Paul R., and Gretchen C. Daily. 1988. Red-naped sapsuckers feeding at willows: possible keystone herbivores. *American Birds* 42: 359–65.

Evans, Julian. 1995. *A Wood of Our Own.* Oxford Univ. Press., Oxford, New York, and Tokyo. This is about a forester who bought a patch of land in England and grew a forest on it.

Fahn, A. 1982. *Plant Anatomy,* 3d ed. Pergamon Press, Oxford, New York, and Toronto.

Ferguson, Ian. 1996. *Sustainable Forest Management.* Oxford Univ. Press, Oxford, New York, and Tokyo.

Foster, Robin B. 1977. *Tachigalia versicolor* is a suicidal neotropical tree. *Nature* 268: 624–26.

Hinckley, George W. 1930. *My Friends the Trees.* Good Will Publishing Company, Hinckley, Maine.

Horn, Henry S. 1971. *The Adaptive Geometry of Trees.* Princeton Univ. Press, Princeton, N.J. Explores trees' geometry from the perspective of photosynthesis.

Johnson, W. C., and C. S. Adkisson. 1986. Airlifting the oaks. *Natural History* (October): 41–46. How jays plant oaks.

Jonas, Gerald. 1993. *North American Trees.* Reader's Digest, Pleasantville, N.Y., and Montreal. An excellent reference book on tree uses, history, and biology.

Jordan, Richard N. 1994. *Trees and People.* Regnery, Washington, D.C.

Kaza, Stephanie. 1993. *The Attentive Heart: Conversations with Trees.* Fawcett Columbine, New York.

Kranz, Harald D., and A. R. Huss. 1996. Molecular evolution of pteridophytes and their relationship to seed plants: evidence from complete 185 rRNA gene sequences. *Plant Systematics and Evolution* 202: 1–11.

Lansky, Mitch. 1992. *Beyond the Beauty Strip.* Tilbury House, Gardiner, Maine. Exposes the damage to the Maine woods by the paper industries and the myths used to justify it.

Little, Charles E. 1995. *The Dying of the Trees.* Penguin Books, New York. An alarming picture of pandemic dying of trees around the world due to industrial pollution, disease, and other causes.

Marchand, Peter J. 1987. *North Woods.* Talman Co., New York.

McLaren, B. E., and R. O. Peterson. 1994. Wolves, moose, and tree rings on Isle Royale. *Science* 266:1555–58.

Meeuse, Bastiaan J. D. 1961. *The Story of Pollination.* Ronald Press, New York.

———. 1996. Caterpillars call out for help! *Butterfly Gardeners' Quarterly* (Spring):3.

Moffett, Mark W. 1997. Tree giants of North America. *National Geographic* 191(1): 44–61.

Oliver, Chadnick D., and Bruce C. Larson. 1996. *Forest Stand Dynamics.* John Wiley, New York.

Peattie, Donald C. 1948. *A Natural History of Trees.* Houghton Mifflin, Boston. This first of many editions includes details about each North American tree species.

Perlin, John. 1989. *A Forest Journey.* Harvard Univ. Press, Cambridge, Mass. Explores the use of wood and forests in relation to the rise and fall of several civilizations.

Perlman, Michael. 1994. *The Power of Trees.* Spring Publications, Dallas, Tex. Explores the psychological power of trees, the relationship between trees and human psyche, and the biophilia of trees.

Peterken, George. 1996. *Natural Woodland.* Cambridge Univ. Press, Cambridge.

Peterson, Merrill A. 1994. Caterpillars and their hired guns. *American Butterflies* (November): 25–28.

Platt, Rutherford. 1942. *This Green World.* Dodd & Mead Co., New York. A popular account of tree and other plant biology.

Raghavendra, A. S., ed. 1991. *Physiology of Trees.* John Wiley, New York.

Robinson, Gordon. 1988. *The Forest and the Trees.* Island Press, Washington, D.C.

Schultz, Jack C. 1983. Tree tactics. *Natural History* 92 (5): 12–25.

Schultz, Jack C., P. J. Northnagle, and I. T. Baldwin. 1982. Seasonal and individual variation in leaf quality of two northern hardwood tree species. *American Journal of Botany* 69: 753–59.

Sheraton, Mimi. 1995. Apples preserved. *Audubon* (November–December): 74–81. The preservation of a genetic apple "library."

Sisler, E., and S. F. Yang. 1984. Ethylene, the gaseous plant hormone. *Bio Science* 34: 234–38.

Smith, D. M., Bruce C. Larson, Mathew J. Kelty, P. Marks, and S. Ashton. 1997. *The Practice of Silviculture.* John Wiley, New York.

Spongberg, Stephen A. 1990. *A Reunion of Trees.* Harvard Univ. Press, Cambridge, Mass. About the trees in the Arnold Arboretum of Harvard University.

Steele, M. L., and P. Smallwood. 1994. What are squirrels hiding? *Natural History* (October): 40–44.

Steudle, Ernst. 1995. Trees under tension. *Nature* 378:663–64. Discussion of current theories of how water rises in trees.

Tainter, Frank H., and F. A. Baker. 1996. *Principles of Forest Pathology.* John Wiley, New York.

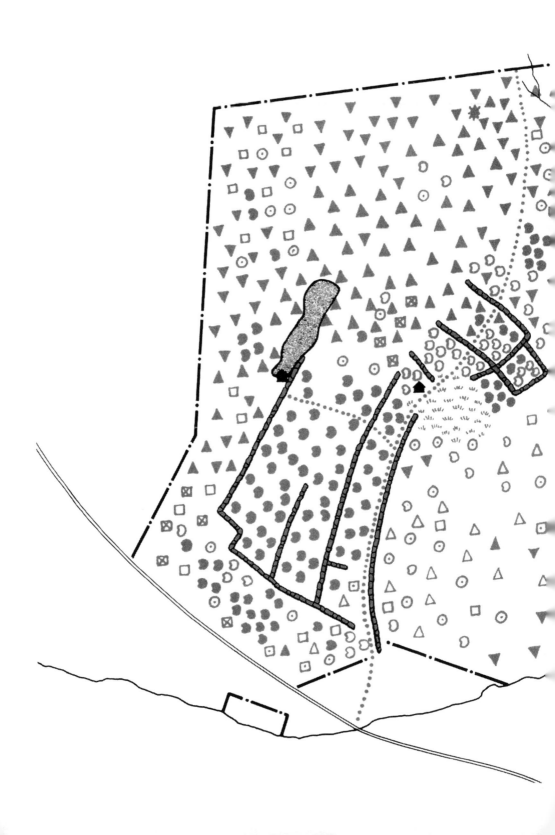

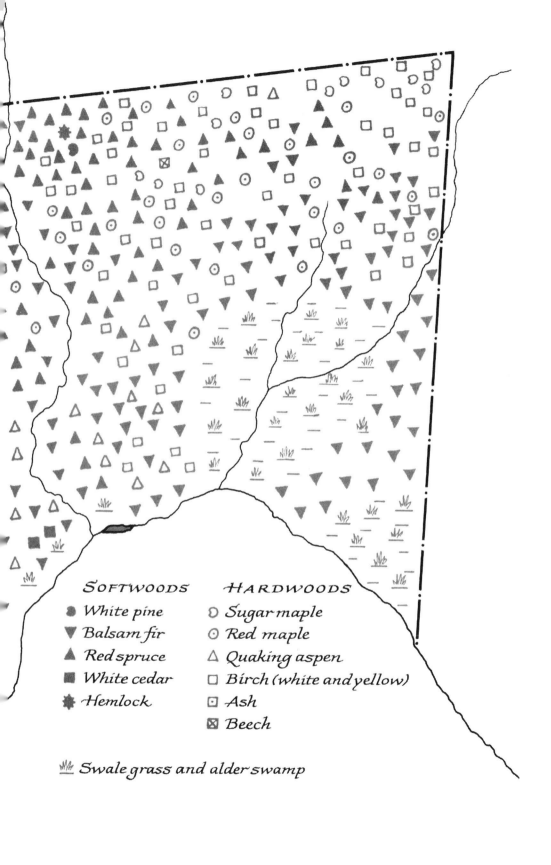

SOFTWOODS

🌑 White pine
▼ Balsam fir
▲ Red spruce
◼ White cedar
✳ Hemlock

HARDWOODS

☽ Sugar maple
⊙ Red maple
△ Quaking aspen
▢ Birch (white and yellow)
⊡ Ash
⊠ Beech

🌿 Swale grass and alder swamp

ecco

Coming May 2004:

THE GEESE OF BEAVER BOG
ISBN 0-06-019745-5 (hardcover)

With a biologist's eyes and a curious nature-lover's soul, Bernd Heinrich set out to the beaver bog adjacent to his house to observe and understand the daily life of Canadian geese—whose routines are as colorful and dramatic as those of their human counterparts.

Books by Bernd Heinrich:

MIND OF THE RAVEN: *Investigations and Adventures with Wolf-Birds*
ISBN 0-06-093063-2 (paperback)

Bernd Heinrich finds himself dreaming of ravens and decides he must get to the truth about this animal reputed to be so intelligent.

"An amazing book by an amazing author. Heinrich has documented a level of intelligence and social sophistication rarely even dreamed to exist in birds."
—Edward O. Wilson

WHY WE RUN: *A Natural History*
ISBN 0-06-095870-7 (paperback)
ORIGINALLY PUBLISHED IN HARDCOVER AS *RACING THE ANTELOPE*

Sinuously weaving evolution, intelligence, and imagination with his own stories of long-distance running, Bernd Heinrich has created an original and provocative work combining the rigors of science with the passion of running.

"A remarkable perspective on how and why running is an integral part of what makes us human." —*UltraRunning* magazine

THE TREES IN MY FOREST
ISBN 0-06-092942-1 (paperback)

Biologist and acclaimed nature writer Bernd Heinrich takes readers on an eye-opening journey through the hidden life of a forest.

"In Heinrich's hands, the lives of trees are as noble and dramatic as the lives of men." —*Washington Post*

Don't miss the next book by your favorite author.
Sign up for AuthorTracker by visiting *www.AuthorTracker.com*.

Available wherever books are sold, or call 1-800-331-3761 to order.
www.harpercollins.com